1st International Conference on Sensing for Industry, Control, Communications and Security Technologies

(ICSICCST 2013)

IOP Conference Series: Materials Science and Engineering Volume 51

Karachi, Pakistan
24 – 26 June 2013

Chief Editor:

A. Shuja Syed

ISBN: 978-1-63439-862-6
ISSN: 1757-8981

Printed from e-media with permission by:

Curran Associates, Inc.
57 Morehouse Lane
Red Hook, NY 12571

Some format issues inherent in the e-media version may also appear in this print version.

Printed by Curran Associates, Inc. (2015)

For permission requests, please contact the Institute of Physics
at the address below.

Institute of Physics
Dirac House, Temple Back
Bristol BS1 6BE UK

Phone: 44 1 17 929 7481
Fax: 44 1 17 920 0979

techtracking@iop.org

Additional copies of this publication are available from:

Curran Associates, Inc.
57 Morehouse Lane
Red Hook, NY 12571 USA
Phone: 845-758-0400
Fax: 845-758-2634
Email: curran@proceedings.com
Web: www.proceedings.com

TABLE OF CONTENTS

IOP Conf. Series: Materials Science and Engineering **51** (2013) 011001 doi:10.1088/1757-899X/51/1/011001

ICSICCST'13

1st International Conference on Sensing for Industry, Control, Communication & Security Technologies

June 24-26, 2013

PREFACE

The 1st International Conference on Sensing for Industry, Control, Communication & Security Technologies (ICSICCST-2013), took place in Karachi, Pakistan, from June 24th – 26th, 2013. It was organized by Indus University, Karachi, in collaboration with H.E.J. Research Institute of Chemistry, University of Karachi, Karachi. More than 80 abstracts were submitted to the conference and were double blind-reviewed by an international scientific committee. The topics of the Conference were:

- Video, Image & Voice Sensing
- Sensing for Industry, Environment, and Health
- Automation and Controls
- Laser Sensors and Systems
- Displays for Innovative Applications
- Emerging Technologies
- Unmanned, Robotic, and Layered Systems
- Sensing for Defense, Homeland Security, and Law Enforcement

The title of the conference, "Sensing for Industry, Control, Communication & Security Technologies" is very apt in capturing the main issues facing the industry of Pakistan and world. We believe the sensing industry, particularly in Pakistan, is currently at a critical juncture of its development. The future of the industry will depend on how the industry players choose to respond to the challenge of global competition and opportunities arising from strong growth in the Asian region for which we are pleased to note that the conference covered a comprehensive spectrum of issues with an international perspective. This will certainly assist industry players to make informed decisions in shaping the future of the industry.

The conference gathered qualified researchers from developed countries like USA, UK, Sweden, Saudi Arabia, China, South Korea and Malaysia etc whose expertise resulting from the research can be drawn-upon to build an exploitable area of new technology that has potential Defense, Homeland Security, and Military applicability. More than 250 researchers / students attend the event and make the event great success as the turnout was 100%. An exceptional line-up of speakers spoke at the occasion.

We want to thank the Organizing Committee, the Institutions and Sponsors supporting the Conference, specially "Centre for Emerging Sciences, Engineering & Technology (CESET), Islamabad", the IOP Publishers and everyone who contributed to the organization of this meeting, for their invaluable efforts to make this event a great success.

IOP Conf. Series: Materials Science and Engineering **51** (2013) 011001 doi:10.1088/1757-899X/51/1/011001

ICSICCST'13

ICSICCST'13

Organized by:

Indus University
Karachi, Pakistan

In Collaboration with:

H.E.J. Research Institute of Chemistry,
University of Karachi, Karachi, Pakistan

Centre for Emerging Sciences,
Engineering & Technology, Islamabad

Institute of Physics

Sponsors:

National ICT R&D Fund

Al-Riaz Green System

Hilton Pharma Ltd

Infrared detectors: Advances, challenges and new technologies

Amir Karim and Jan Y Andersson
Acreo Swedish ICT AB., Dept of Nanoelectronics, Electrum 236, 16440 Kista, Sweden

amir.karim@acreo.se

Abstract. Human knowledge of infrared (IR) radiation is about 200 years old. However it was in the late 20th century that we developed a wide range of smart technologies for detection and started to take advantage for our benefit. Today IR detector technology is in its 3rd generation and comes with challenging demands. Based on the propagation of IR radiation through free space it is divided into different transmission windows. The most interesting for thermal imaging are the mid-wave IR (MWIR) and the long-wave IR (LW IR). Infrared detectors for thermal imaging have a number of applications in industry, security, search & rescue, surveillance, medicine, research, meteorology, climatology and astronomy. Currently high-performance IR imaging technology is mainly based on epitaxially grown structures of the small-bandgap bulk alloy mercury-cadmium-telluride (MCT), indium antimonide (InSb) and GaAs based quantum-well infrared photodetectors (QWIPs), depending on the application and wavelength range. However, they operate at low temperatures requiring costly and bulky cryogenic systems. In addition there is always a need for better performance, which generates possibilities for developing new technologies. Some emerging technologies are quantum dot infrared photodetectors (QDIPs), type-II strained layer super-lattice, and QDIPs with type-II band alignment. In this report a brief review of the current and new technologies for high performance IR detectors, will be presented.

1. Introduction

This report presents a brief review of IR detectors with a focus on the challenges faced by the current IR detector technologies and demands for future. The discussion is also focused on the demands for next generation IR detectors, with a short overview about emerging technologies. However before starting the discussion about IR detectors it is worthwhile to give a brief overview of the IR radiation itself.

2. A brief overview of IR

IR radiation is the most common form of electromagnetic radiation lying between the visible and microwave part of the spectrum. In terms of wavelength it is between ~0.76μm up to 1000μm. The first discovery of IR was made by William Herschel in 1800 through a simple experiment [1]. However a more clear understanding of it was developed in 1900 through Plank's law, which is a pioneer result of modern physics and quantum theory. Every physical object in the universe spontaneously emits radiation with a broad range of wavelengths. The peak emission wavelength corresponds to the equilibrium temperature of the object. The spectral radiant emittance follows Plank's relation as shown in figure 1. The peak emission of objects at room temperature (~300 K) is at

10 μm. Whereas, the surface of the sun (~6000 K) has a peak emission in the visible range containing a great amount of IR and ultraviolet (UV) radiation.

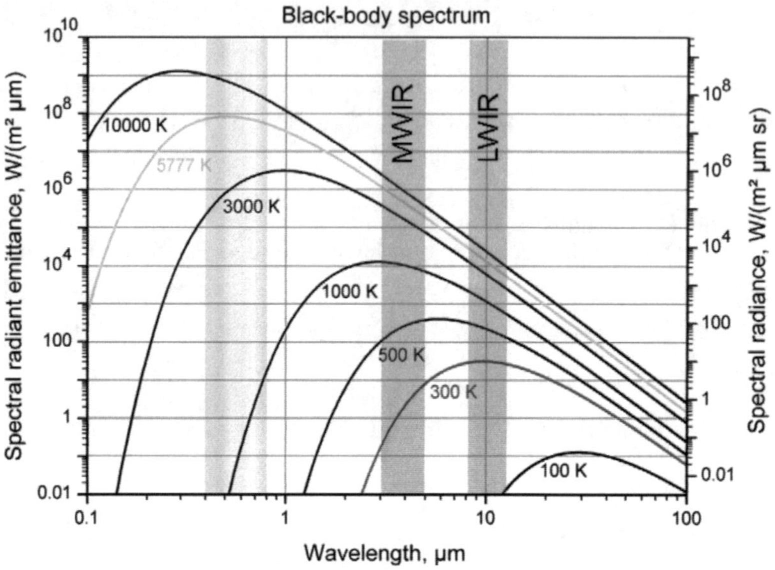

Figure 1. Spectral emittance of objects at given equilibrium temperatures.

Upon transmission through atmosphere, IR radiation is absorbed by different chemical compounds found in the atmosphere, for example H_2O and CO_2. There may exist more than one definitions to subdivide the IR radiation spectrum. Under standard conditions, there are two interesting minimum absorption windows,3-5 μm and 8-12 μm, which are referred as mid-wave IR (MWIR) and long-wave (LWIR), respectively.

3. Applications

Heat sensors were developed even before the full understanding of IR. However it was from the mid-20th century when a wide range of IR detector technologies were developed and exploited for our benefits. IR detectors for thermal imaging have numerous applications. A few of the important applications are mentioned below.

IR radiation is less absorbed and scattered in the atmosphere compared to the visible light and provides important information about objects, for example their temperature, geometry, composition, location in space, and atmosphere. IR imaging has contributed a lot in astronomy to understand the universe [2, 3].NASA space telescopes are equipped with high performance IR imaging systems, which have provided tremendous information of galaxies and star clusters with details that were previously un seen with visible-light imaging [4].

There are also huge applications in military [5]. Infect initially the IR technology was heavily funded and developed for military applications. Today many military aircrafts are equipped with high performance IR cameras for scanning the battlefield in poor visibility situations. Apart from that there are also ground based night vision systems used by military. There are also IR based missile tracking systems.

Other applications of thermal imagine are in, medical [6], surveillance, search and rescue, meteorology, and climatology.

4. Types of IR detectors

There are two main types of IR detectors, thermal and photonic. Thermal detectors operate based on detecting the thermal effects of the incident IR radiation. The thermal sensing mechanism occurs through many temperature dependent phenomena. Examples of thermal detectors are thermocouples, thermopiles and bolometers. Bolometers are the most common type of thermal detectors, which are based on changes in resistance of the heat sensing material due to the incident IR radiation. Thermocouples and thermopiles use the thermoelectric effect. On the other hand in photonic detectors the incident IR radiation causes intrinsic or extrinsic electronic excitations. Photonic IR detectors can be photoconductive or photovoltaic. In photoconductors the resistivity of the detector element is monitored. Whereas photovoltaic detectors are basically pn diodes and a photoelectric current is generated upon the IR illumination.

From the IR market point of view one can talk about mainly three parameters, volume, performance and cost. The thermal detectors are basically high volume but low performance and low cost technology. However the photonic detectors are low volume but high performance and high cost market. One of the big factors that contribute in their cost is the cooling system. Photonic detectors need to operate at low temperatures. Most of them operate at liquid nitrogen temperature (77K), or ~100K, depending on the wavelength and the detector material. Whereas thermal detectors operate at room temperature. There is another range of photonic detectors with intermediate volume, performance and cost. This is the region where a lot of new competing technologies are emerging.

5. IR detector technologies

5.1. Early detector technologies

The early detector technologies were developed in1940's mainly based on compound semiconductors. The first practical IR detector was based on lead sulfide (PbS) for detection up to 3μm wavelength range [7]. In late 40's this detection wavelength was extended up to 5μm using lead selenide (PbSe), lead telluride (PbTe), and indium antimonide (InSb). Activities ongroup III/V, IV/IV and II/VI semiconductor alloys started in 1950's which lead to the development of new semiconductor alloy materials. Mercury-cadmium-telluride (MCT) was one of them, which covers a broad range of IR wavelengths.

5.2. Current technologies and challenges

The current IR technology is mainly based on MCT, which is a narrow and direct bandgap zincblende II-VI ternary alloy of CdTe and HgTe [8, 9]. It has a tunable bandgap spanning the shortwave IR to the very long wave IR regions. The amount of cadmium (Cd) in the alloy can be chosen to tune the optical absorption cutoff wavelength of the material to the desired infrared wavelength. CdTe is a semiconductor with a bandgap of approximately 1.5 eV at room temperature and HgTe is a semimetal with zero bandgap energy. Hence by mixing these two materials one can obtain any bandgap energy between 0 and 1.5 eV. The IR detection mechanism is based on interband transitions between conduction band and valance band, as shown in figure 2.

MCT is a high cost technology mainly due to the fact that it is grown on ZnCdTe. Apart from that there are also uniformity issues for long wave region, where the detection wavelength is more sensitive to compositional variations. Other than compositional inaccuracies, high Auger recombination rates lead to large dark currents. Nevertheless it is currently the dominating IR detector technology for applications with multiple wavelength ranges, where cost is not an issue.

Another existing IR technology is quantum-well infrared photodetectors (QWIPs) [10, 11]. The detector material is based on AlGaAs/GaAs QWs grown on GaAs substrates. The detection mechanism is dependent on intersubband transitions as shown in figure 2, where the excitation of electrons from QW ground state to excited state is shown. It can also cover a wide range of wavelengths. It is a lower cost technology compared to MCT, with high uniformity and best for large format arrays. But it suffers from low quantum efficiency and operating temperature is also lower than MCT. There is also polarization dependence of the incident IR radiation, which requires fabrication of special grating structures on top of each pixel. Acreo Swedish ICT has contributed to a large extent in developing QWIPs for several years and QWIP based high-performance detector arrays [12, 13, 14].

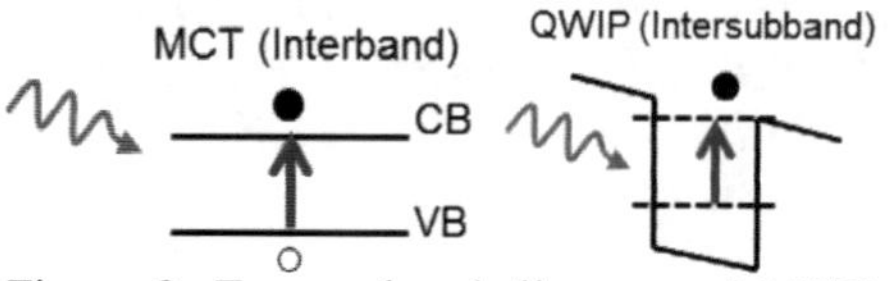

Figure 2. Energy band diagram of MCT and QWIP showing electronic transition.

6. New IR technologies

The motivation for the current and new IR detector technology developments is to look for higher operating temperatures, higher efficiency and lower cost. The continuous quest for larger detectivities, lower operating power and higher operating temperatures necessitates new exotic materials. Two of these new technologies are quantum dot based infrared photodetectors (QDIPs) [15-17] and type-II strained layer superlattice [18-20]. Apart from that there is another technology with a new type of QDIP scheme based on QDs with type-II band alignment [21-24]. These are the technologies we at Acreo Swedish ICT, Sweden, have been involved for some years [17, 20, 22-25]. In this report QDIPs and type-II superlattice are briefly introduced and a more detailed description about QDIPs with type-II band alignment is given.

6.1. QDIPs

Quantum dots emerged in 1970's as nanometre size features of one semiconductor material in another. These are zero dimensional semiconductor structures with unique optical and electronic properties as a result of quantum confinement. The development of QDIP structures was actually stimulated after the success of QWIPs. They are similar as QWIPs but with extra advantages due to the three dimensional confinement in the QDs. The detection mechanism in QDIPs is also based on intersubband transitions between the quantized energy levels of the dots and continues states [15-17].

QDIP has emerged as a promising technology for third-generation imaging system and there has been a fast development of this technology over the last decade. The material system of QDIPs is mainly based on InAs QDs grown on GaAs substrate. However, several other heterostructure designs have also been investigated for use as IR photodetectors [26-28].

One of the biggest anticipated advantage of QDIPs is the possibility of normal incident detection, unlike in QWIPs, therefore removing the fabrication of grating coupler on imaging arrays. Other expected benefits of QDIP technology are longer carrier life time compared to QWIPs leading to lower dark currents, higher operating temperatures, multicolour detection, and lower cost compared to MCT.

QDIPs have already demonstrated imaging in MWIR and LWIR range, however currently it suffers from lower absorption quantum efficiency in comparison with the interband type of photodetectors. Nevertheless, there might be applications where incident photon flux is high. For example, in the photon-rich terrestrial regime QDIPs can achieve similar performance as the interband type of photodetectors due to very low dark current levels.

6.2. Type-II superlattice

The type-II superlattice strained layer structures have recently emerged as a promising material for high-performance IR photodetectors in MWIR as well as LWIR range [18-20]. This material system consists of a few monolayers (MLs) of InAs and a few MLs of GaSb, which are repeated in a number of periods. These layers are basically grown on GaSb (100) substrates. Although the lattice mismatch between InAs and GaSb is less than one percent, InAs is tensile strained. These quantum structures are characterized by a broken bandgap with type-II band alignment leading to spatially indirect transitions between hole states localized in the GaSb layers and delocalized electronic states in the InAs layers, as shown in the schematic energy band diagram of figure 3. The effective band gap of these structures can be tailored from 0.3 eV to values below 0.1 eV by varying the thickness and composition of the superlattice layers. The detector devices based on this material system are photovoltaic. There has been a lot of work on device development leading to exotic type of device structures in order to optimize detector performance.

Dark current is an important issue in these devices. The dark current has two components, bulk leakage and surface leakage. The bulk component depends on the material quality whereas the surface component is a result of device fabrication. Over the years the material quality has been greatly improved due to advancements in growth techniques, particularly using the growth technique of MBE. Minimization of surface leakage is essential in device fabrication. A proper etching and sidewall passivation can improve the situation greatly. A good understanding of mesa sidewall and passivation interface is essential to control surface leakage [20]. For cost reduction it would also be beneficial to realize InAs/GaSb SL growth on GaAs substrates.

6.3. QDIPs with type-II band

This is another new technology for IR detector applications. Although there has not been much work done on this area, it offers some very interesting advantages for realizing IR detectors [21-24]. It is also based on QDs but with type-II band alignment. In this case the conduction band of the matrix material is lower in energy than the valance band of the dot material. The energy band alignment in this material system is such that the holes are confined in the quantized energy states of dots and electrons are free to move in the conduction band of the matrix material. The detection mechanism in this system is based on the spatially indirect band-to-band transitions between QDs and the bulk material, as shown in figure 3. Hence it combines all the predicted advantages of QDIPs, as mentioned in section 6.1, with the expectation of stronger interband absorptions as compared to intersubband absorptions.

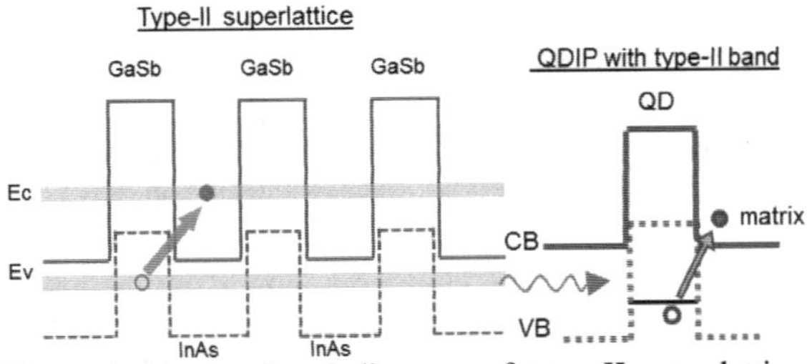

Figure 3. Energy band diagram of type-II superlattice and QDIPs with type-II band alignment, showing electronic transition.

6.3.1. Material system

There are more than one possible QD material systems with type-II band alignment, for example GaAs QDs on InAs [22] and InSb QDs in InAs [29]. Both of these material systems have similar lattice mismatches of about 7%, which leads to the formation of self-assembled QD after the critical thickness of GaAs or InSb deposition. At room temperature GaAs has a bandgap of ~1.4eV and InAs has bandgap of ~0.35eV. The lattice constant of GaAs is smaller than InAs, hence GaAs dots on InAs are tensely strained. On the other hand InSb with a room temperature bandgap of ~0.17eV,has a larger lattice constant than InAs and InSb QDs on InAs are compressively strained.

Since the confinement in the dots is dependent on QD size and compositions, therefore the corresponding interband electronic transition can be tuned by controlling the dot size and composition.

It is evident that for larger dot sizes the quantized energy states of dots would shift closer to the matrix (InAs) conduction band leading to a red shift of the absorption wavelength, and vice versa. A possibility of composition variation for wavelength tuning in case of GaAs dots is discussed in reference 22 by incorporating Al and Sb in the dots. Whereas, in InSb dots the incorporation of Ga influences the dot to matrix transition in order to achieve LWIR absorption, as reported in reference 24. Apart from that there are a number of growth parameters that influence dot formation and as a result the transition wavelength. For example in case of material grown using the technique of metal organic vapour phase epitaxy (MOVPE) growth development is performed by optimizing growth temperature, group V and group III molar ratios, growth rate, and amount of material deposited.

6.3.2. Detector device structure

The detector devices are fabricated by well-known III-V device fabrication technology using wet and dry etching followed by side wall chemical polishing and metallization. Surface leakage is an important issue in these devices as well. Therefore a proper sidewall passivation is required to reduce dark currents generated by surface leakage. Apart from that there is also need for device design optimization to develop barrier structures.

Nevertheless, the device structure is pin diode type with multiple QD layers embedded in the intrinsic InAs layer between the p and n doped layer. Schematic view of a fabricated single pixel device is shown in figure 4. The IR radiation is incident from the top which is absorbed by the dots generating electron hole pairs. The photo generated electrons are free to move in the conduction band of matrix material to contribute in photocurrent, however holes are bound in the quantized states of dots. This is the reason for thermally activated photoresponse reported in In(Ga)Sb/InAs QD based IR photodetectors [24].

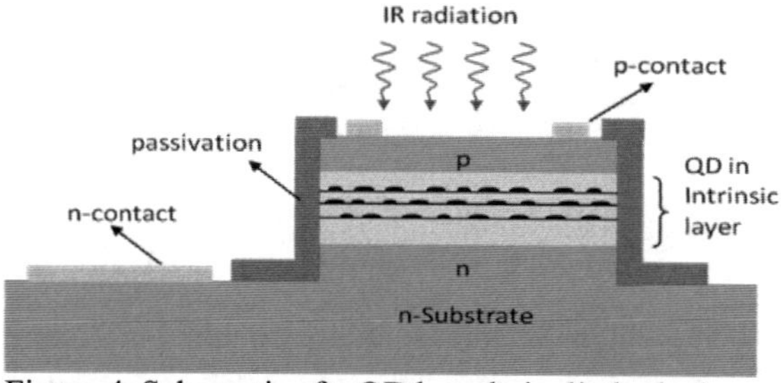

Figure 4. Schematic of a QD based pin diode device.

Among all the new emerging technologies, QDIPs with type-II band is the least evolved. There has not been much work done for photodetectors application, although there has been more work performed using this approach for IR light emitters. Currently the most severe issue in the material technology is the low quantum efficiency.

7. Conclusions and future outlook

High-performance IR detector technology is entering into its 3rd generation with high demands on detection efficiency, operating temperatures and particularly ability to perform multicolour detection. A few of the important emerging IR detector technologies were briefly discussed in this report. However none of these new technologies are fully mature and commercialised yet, although some are more and some are less developed.

QDIP based IR photodetectors are anticipated to have advantages of low dark current, high operating temperature, normal incidence and multicolour detection. There has been quite some work performed on QDIPs and many of these advantages have been demonstrated but this technology is suffering from low quantum efficiencies. Nevertheless there might be applications where incoming photon flux is high.

QDIPs with type-II band alignment are expected to have all the advantages offered by QDs as well as higher detections efficiencies because of the band-to-band transitions, unlike in conventional QDIPs. However the quantum efficiency values experimentally demonstrated so far are still very low. The dark current values reported are also not yet close to theoretically anticipated. Among all the three emerging technologies discussed in this report QDIPs with type-II band is the least developed one. The most challenging parameters are full control of the self-assembled QDs, in terms of their size, compositions and material quality. Apart from that device processing and device design also need to be improved to reduce dark currents. Efforts are also required to address the thermal activation of photoresponse, which is a result of holes trapped in the dots after photo excitation.

Type-II strain layer superlattice is the most developed among the new IR detector technologies. It has a flexible bandgap like MCT, depending on the SL thickness and composition. Successful demonstrations of MWIR imaging arrays operating at 120K are reported with efficiencies close to 40%. Reports of single pixel LWIR detectors have also been made based on type-II superlattice. The challenges in this technology are SL interfaces and device passivation. Large developments have been made in terms of material growth particularly applying the technique if MBE. Since this technology is based on number of very thin QWs on top of each other, a good understanding and control of SL interfaces is essential. Apart from that growth of these SL on GaAs would also be beneficial for reducing cost.

A vision for future of IR detector technology is photodetectors integrated with powerful hardware performing smart algorithms. We also may see arrays with each pixel sensing the whole IR spectrum. This may lead to bio-inspired sensing and perhaps fabrication of a complete IR retina.

References

[1] "Sensing the Invisible: The Herschel Experiment", http://sciencenetlinks.com/lessons/sensing-the-invisible-the-herschel-experiment/.

[2] J.W. Beletic, R. Blank, , D. Gulbransen, , D. Lee, M. Loose, E.C. Piquette, , T. Sprafke, W.E. Tennant, M. Zandian, J. Zino, "Teledyne Imaging Sensor:Infrared imaging technologiesfor Astronomy & Civil Space", Infrared SPIE Paper 7021-20 (2008).

[3] R.D. Gehrza, E.E. Becklinb, I. de Paterc, D.F. Lesterd, T.L. Roellige, C.E. Woodward, "A new window on the cosmos: The Stratospheric Observatory for Infrared Astronomy (SOFIA)", Advances in Space Research 44, 413-431, (2009).

[4] IR processing and analysis center NASA http://www.ipac.caltech.edu/gallery.

[5] Latika Becker, "Influence of IR Sensor Technology on the Military and Civil Defense", Proc. of SPIE Vol. 6127, (2006).

[6] E.F. Ring and Ammer K, "Infrared thermal imaging in medicine", Physiol. Meas. 33, R33-46, (2012).

[7] Carlo Corsi, "History highlights and future trends of infrared sensors", Journal of modern optics, Vol. 57, No. 18, 1663-1686, (2010).

[8] W.D. Lawson, S. Nielson, E.H. Putley and A.S. Young J. Phys. Chem. Solids 9, 325–9, (1959)

[9] A. Rogalski, Rep. Prog. Phys. 68, 2267–2336, (2005).

[10] H.C Liu, C.Y. Song, A. Shen, M. Gao, Z.R. Wasilewski, and M. Buchanan, Appl. Phys. Lett. 77, 2437-2439, (2000).

[11] L. Shkedy, T. Markovitz, Z. Calahorra, I. Hirsh, and I. Shtrichman, Optical Engineering 50(6), (2011).

[12] J. Y. Andersson and L. Lundqvist, "Near-unity quantum efficiency of AlGaAs/GaAs quantum well infrared detectors using a waveguide with a doubly periodic grating coupler", Appl.Phys. Lett. 59 p.857-859, (1991).

[13] J. Y. Andersson, "Dark current mechanisms and conditions of background radiation limitation of n-doped AlGaAs/GaAs quantum-well infrared detectors", J. Appl. Phys. 78, pp. 6298-6304, (1993).

[14] H. Martijn, S. Smuk, C. Asplund, H. Malm, A. Gromov, J. Alverbro, H. Bleichner, Proc. of

SPIE Vol. 6542, 65420V, (2007).

[15] Gunapala S.D., Bandara S.V., Hill C.J., Ting D.Z., Liu J.K., Rafol S.B., Blazejewski E.R., Mumolo J.M., Keo S.A., Krishna S., Chang Y.-C., and Shott C.A., IEEE Journal of Quantum Electronics, 43(3), (2007).

[16] David A. Ramirez, Jiayi Shao, Majeed M. Hayat, and Sanjay Krishna, "Midwave infrared quantum dot avalanche photodiode", APL.**97**, 221106, (2010).

[17] S. Hellström, Z.-H. Chen, Y. Fu, M. Qiu, R. Soltanmoradi, Q. Wang and J. Y. Andersson, "Increased photocurrent in quantum dot infrared photodetector by subwavelength hole array in metal thin film", Appl. Phys. Lett. 96, 231110 (2010).

[18] M. Razeghi, E.K. Huang, B.M. Nguyen, S.A. Pour and P.Y. Delaunay, Proc. of SPIE Vol. 7660, 76601F (2010).

[19] Eric A. DeCuir, Jr., Gregory P. Meissner, Priyalal S. Wijewarnasuriya, Nutan Gautam, Sanjay Krishna, Nibir K. Dhar, Roger E. Welser, Ashok K. Sood, "Long-wave type-II superlattice detectors with unipolar electron and hole barriers", Opt. Eng. 51(12), 124001 (Dec 03, 2012).

[20] Amir Karim, Rickard Marcks von Würtemberg, Carl Asplund, Hedda Malm, Jan Andersson, Elena Plis, Sanjay Krishna, " Characterization of InAs/GaSb type-II superlattice photodiodes for mid-wave IR with different mesa sidewall passivation schemes", Phys. Status Solidi C, 9: 1690–1692, (2012).

[21] C.J. Hill, A. Soibel, S.A. Keo, J.M. Mumolo, D.Z. Ting and S.D. Gunapala, Electronics letters 46, 1286 (2010).

[22] O. Gustafsson, J. Berggren, U. Ekenberg, A. Hallén, M. Hammar, L. Höglund, A. Karim, B. Noharet, Q. Wang, A. Gromov, S. Almqvist, A. Zhang, S. Junique, J.Y. Andersson, C. Asplund, R. Marcks von Würtemberg, H. Malm, H. Martijn, Infrared Physics & Technology 54, 287–291, (2011).

[23] A. Karim, O Gustafsson, L. Hussain, Q. Wang, N. Noharet, M. Hammar, J. Anderson, J. Song, Proc. SPIE 8439 (2012).

[24] A. Karim, O. Gustafsson, S. Savage, Q. Wang, S. Almqvist, C. Asplund, M. Hammar, J.Y. Andersson, "In (Ga) Sb/InAs quantum dot based IR photodetectors with thermally activated photoresponse", SPIE Defense, Security, and Sensing, 870434-870434-9 (2013).

[25] J. Y. Andersson, L. Höglund, B. Noharet, Q. Wang, P. Ericsson, S. Wissmar, C. Asplund, H. Malm, H. Martijn, M. Hammar, O. Gustafsson, S. Hellström, H. Radamson, P. O Holtz, "Quantum structure based infrared detector research and development within Acreo's centre of excellence IMAGIC", Infrared Physics & Technology 53, 227–230, (2010).

[26] P. Bhattacharya, Z. Mi, Proc.IEEE95, 1723, (2007).

[27] J.C. Campbell, A. Madhukar, Proc.IEEE95, 1815, (2007).

[28] L. Höglund, P. O. Holtz, H. Pettersson, C. Asplund, Q. Wang, H. Malm, S. Almqvist, E. Petrini and J. Y. Andersson, Appl. Phys. Lett. 94, 053503, (2009).

[29] O. Gustafsson, A. Karim, J. Berggren, Q. Wang, C. Reuterskiöld-Hedlund, C. Ernerheim-Jokumsen, M. Soldemo, J. Weissenrieder, S. Persson, S. Almqvist, U. Ekenberg, B. Noharet, C. Asplund, M. Göthelid, J.Y. Andersson, M. Hammar, "Photoluminescence and photoresponse from InSb/InAs-based quantum dot structures" Optics Express, 20 (19), 21264-21271 (2012).

IOP Conf. Series: Materials Science and Engineering **51** (2013) 012002 doi:10.1088/1757-899X/51/1/012002

A Comparison of Generative and Discriminative Appliance Recognition Models for Load Monitoring

Ahmed Zoha #1, **Muhammad Ali Imran** 1, **Alexander Gluhak** 1, **Michele Nati** 1

1Centre for Communication Systems Research, University of Surrey, GU2 7XH, Guildford, UK

E-mail: `a.zoha@surrey.ac.uk`

Abstract. Appliance-level Load Monitoring (ALM) is essential, not only to optimize energy utilization, but also to promote energy awareness amongst consumers through real-time feedback mechanisms. Non-intrusive load monitoring is an attractive method to perform ALM that allows tracking of appliance states within the aggregated power measurements. It makes use of generative and discriminative machine learning models to perform load identification. However, particularly for low-power appliances, these algorithms achieve sub-optimal performance in a real world environment due to ambiguous overlapping of appliance power features. In our work, we report a performance comparison of generative and discriminative Appliance Recognition (AR) models for binary and multi-state appliance operations. Furthermore, it has been shown through experimental evaluations that a significant performance improvement in AR can be achieved if we make use of acoustic information generated as a by-product of appliance activity. We demonstrate that our a discriminative model FF-AR trained using a hybrid feature set which is a catenation of audio and power features improves the multi-state AR accuracy up to 10 %, in comparison to a generative FHMM-AR model.

1. Introduction

It is well known that due to growing energy demands, specifically in the last few decades, energy conservation is becoming challenging. This is particularly true for residential spaces and within offices that accounts for almost *40* % of total European Union (EU) energy consumption, generating almost *36* % of Green House Gas (GHG) emission. Moreover, it has been predicted that global energy demands will double by the end of 2030 [1], which leads to increased energy prices along with its negative implications (i.e., CO_2 emissions) on the environment. This substantial growth in energy consumption directly impacts the economy of a country. In this context, information and communication technologies (ICT), more specifically Internet of Things (IoT), can prove to be important enablers of smart solutions that reduce the overall energy consumption. A detailed review [2] of more than 60 feedback studies suggest that maximum energy saving can be achieved using direct feedback mechanisms (i.e., real-time appliance level consumption information) as opposed to indirect feedback mechanisms (i.e., monthly bills, weekly advice on energy usage). To achieve this real-time energy information we see a large scale deployment of smart meters by the governments of UK and USA. However, these meters provide little information on the breakdown of energy spent. This has motivated research efforts to develop Appliance-level Load Monitoring (ALM) methods to achieve fine-grained energy

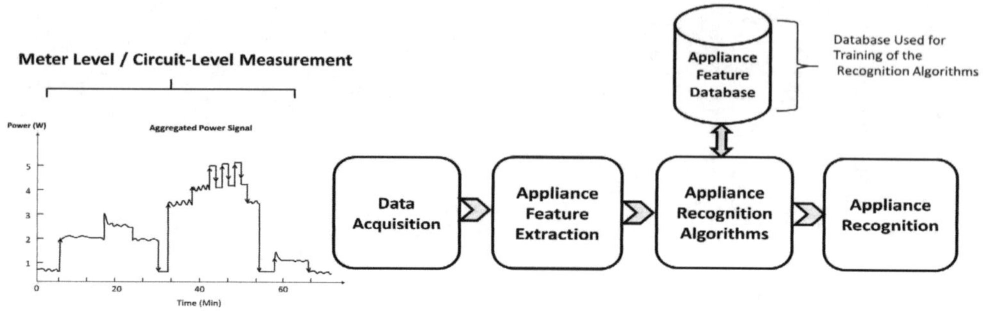

Figure 1. A General Framework for Non-Intrusive Load Monitoring

monitoring. The goal of the detailed energy information acquired from these ALM methods is to make the invisible flow of energy visible to the consumers. This has a direct impact on the behavior of the users, as mostly they are unaware of their energy wastage behavior. Moreover, energy-awareness is important so that users can differentiate between different energy providers, and develop their own strategies for energy saving. Consumers are able to cut down their power usage through knowing what devices are often used purposelessly. From a smart-grid perspective, demand side management could be supplemented by having an analysis of appliance level energy usage information which provides a better insight into power utilization and facilitates formation of optimum resource allocation strategies. ALM is not only a pre-requisite for precise energy feedback, but it is equally applicable for other applications such as fault detection, monitoring and security as well as for the automated energy management systems.

The remainder of the paper is organized as follows. In the next section, we review related research work, highlighting the limitation of current ALM approaches. We present an overview of our generative and discriminative AR models in Section 3. The experimental setup and performance comparison of our proposed models have been reported in Section 4. Finally, we provide conclusion and future work in Section 5.

2. Load Monitoring Approaches

Research efforts to achieve finer granularity of energy-consumption statistics has led to the development of intrusive and Non-intrusive load monitoring (NILM) techniques. A major drawback of intrusive load monitoring system is that, in order to measure appliance-level energy consumption, multiple sensors needs to be installed on each target appliance. Though we can achieve a accurate measure of device energy usage through this approach but high installation complexity, cost, as well as sensor calibration and data aggregation are outstanding issues that will not favor the use of this technique. NILM, on the other hand uses aggregated power measurements acquired from a single point (i.e., meter or circuit-level) and employ feature extraction methods to extract device features. These features are further used to model AR learning algorithms to disaggregate device specific power usage patterns. Lately, NILM methods have gained a lot of attention in the research community, because of its ability to overcome some of the major challenges faced by intrusive load monitoring systems, as highlighted above.

A general framework for NILM is shown in Figure (1), which involves feature extraction followed by an application of AR algorithm. A change in the aggregated power measurements is observed whenever an appliance switches its state (i.e., ON to OFF), which can be further characterized into steady state and transient events, as proposed by Hart [3]. Accordingly, steady-state and transient-state feature extraction methods are being developed to create a unique signature that will identify each appliance. These features are monitored for tracking load operations in the power measurements and further classified to a particular load using intelligent classification algorithms. The classification algorithms need to be trained by providing labeled examples for training, which first requires extraction of device signatures independently to form an appliance feature database. The AR algorithms reported in literature are based on either pattern recognition [4, 5] or optimization based approaches [6, 7]. Comparatively, pattern recognition approaches that are based on either discriminative or generative machine learning models have shown to perform better than optimization approaches [8]. However, load identification is highly dependent on extracted appliance signatures particularly for discriminative AR models.

The choice of using either steady-state or transient signatures has its own advantages and disadvantages. Steady-state load signatures represent the steady-state behavior of the appliance; therefore it can be obtained from the low-frequency sampling of current and voltage waveforms. High-power appliances exhibit distinct steady-state signatures, however devices with low-power consumption profile are difficult to disaggregate/recognize from the aggregated load measurements due to overlapping steady-state features. In contrast, transient signatures require high sampling of power waveforms to extract shape, size, duration and high order harmonics features to characterize an appliance state transition event. Although, transient features in conjunction with steady-state features have shown to improve the overall disaggregation accuracy of AR models, it requires costly hardware, in addition they are sensitive to the wiring architecture and demand excessive training of the algorithms. Moreover, if the system needs to be envisioned for residential spaces, current smart meters are only able to provide data at a low frequency resolution. Nevertheless, most of the existing solutions reported in literature [8] achieve sub-optimal performance in a real-world environment due to following reasons:

- Firstly, most research work is focused on identifying HVAC appliances whereas the presence of low-power appliance loads is often neglected. It is still a challenge for traditional NILM solutions to identify low-power appliances with high accuracy. In our work, we show that using circuit-level power measurements it is possible to accurately discern low-power appliances which are particularly of interest to profile energy consumption of the users within offices.
- Secondly, it is often the case that only binary state (i.e., ON or OFF) operation of the appliances are considered for evaluating the performance of AR algorithms. However, in a real-world setting, appliances often operate in multiple states. Therefore, we have considered binary as well as multi-state operation of the devices in our experimental evaluations, along with adequate feature set selection for optimal classification.
- Thirdly, as highlighted above, AR models trained on steady-state features alone achieve low recognition accuracy, particularly for low-power appliances due to similarity of the appliance features. We propose a discriminative AR model that make use of auxiliary acoustic information associated with the appliance operational state, in order to overcome the ambiguous steady-state feature overlap. Moreover, a comparison has been provided with a generative AR model that employs steady-state features alone.

In the next section, we provide an overview of our discriminative and generative AR models.

3. Our Proposed AR Models

3.1. Factorial Hidden Markov Model Based Appliance Recognition Model (FHMM-AR)

In [9] we propose a generative AR model based on Factorial Hidden Markov Model (FHMM-AR) in order to perform load monitoring in the aggregated power measurements. The FHMM-AR tries to model the joint distribution of hidden appliance states and the power measurements. A structured variational approximation method is adopted for learning, whereas inference is

Table 1. Features used to train our proposed AR models *

Features	Description
Features for Generative Model f_G	$f_G = [P,\ Q,\ \Delta P,\ PF, P_{std}]$, where P and Q are the active and reactive power, PF is the power factor, and P_{std} is the standard deviation of power values within each frame
Audio Features f_A	13 coefficients of Frequency Filtered Band Energies (FFBE's) obtained by applying a second order filter $H(z) = z - z^{-1}$at the output of mel-scaled filter bands
Features for Discriminative Model f_D	$f_D = \{f_A + f_{Dp}\}$ whereas f_{Dp} is the catenation of $[P,\ Q,\ I_{RMS},\ V_{RMS},\ PhaseAngle]$

* We refer the reader to [9] and [10] for a detailed description of the feature sets mentioned in the Table 1

performed using standard Baum-Welch procedure. The feature vector f_G as shown in Table (1) is selected after performing optimal feature selection, and is used to train the final FHMM-AR model. The complete algorithmic details of FHMM and results of feature selection evaluations have been provided in [9].

3.2. Feature Fused Appliance Recognition Model (FF-AR)
In [10] we propose a discriminative AR model based on Support Vector Machines (SVM), whereas the selection of optimal feature set and comparison with other machine learning models have also been presented. It has been shown that proposed model when trained using steady-state features alone achieves low recognition accuracy, due to their ambiguous overlapping in the feature space. To overcome this problem, a solution is proposed that exploits the complimentary acoustic information which is generated as a byproduct of appliance operation. A number of audio features have been extracted and optimal feature selection is performed along with the power features to jointly characterize the audio and power events. It has been demonstrated that the same discriminative model trained using a fused feature vector which is a catenation of f_A and f_{Dp} as shown in Table (1) improves the discriminative ability of classifier. In the text to follow, we refer this model as FF-AR model which has been compared against the FHMM-AR model in this paper, for recognizing binary and multi-state operation of the appliances.

4. Experimental Evaluation
4.1. Experimental Setup
The experimental setup as shown in Figure (2) is similar to the one reported in [9]. To train and test the appliance models, data is gathered from a target user desk via a smart power outlet. This smart outlet has a inbuilt energy meter co-located at user work desk so that 5 target appliances including Work Station (WS), LCD, Laptop, Desk Lamp and a fan can be attached to it via multi-socket. In addition, it has a audio sensor interfaced with it to monitor the acoustic activity of the environment, so that appliance acoustics (i.e. Keyboard typing sound and Fan Operational Sound) could also be captured and processed. This smart power outlet acts as a circuit-level monitoring device which logs aggregated power readings as well as audio measurements of all the appliances at $3Hz$ and $16KHz$, respectively. The acquired data is transmitted to a selected aggregation point (sink), whereas the sink is further connected to a Management Gateway (GW) which reports the data to a central server, so that it can be stored in a database. The monitoring station queries the database periodically to acquire multi-modal measurements for off-line processing.

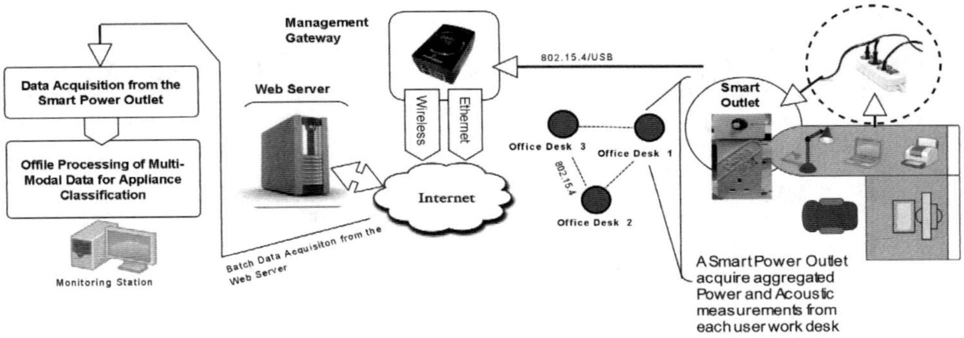

Figure 2. Experimental Test Bed Architecture for Appliance Monitoring at user work desk

Table 2. Disaggregation Accuracy of FHMM based AR Model versus Feature Fused AR Model

Test No	Appliance Combinations	FHMM-AR		FF-AR	
		Binary	Multi-State	Binary	Multi-State
1	WS, LCD	0.987	0.956	0.960	0.950
2	Laptop,Lamp	0.980	0.961	0.950	0.930
3	Lamp, Fan	0.976	0.87	0.940	0.920
4	Laptop, Fan	0.862	0.702	0.943	0.90
5	WS, LCD,Lamp	0.941	0.890	0.950	0.923
6	WS, Lamp,Laptop	0.931	0.91	0.930	0.910
7	WS, LCD,Lamp,Laptop	0.90	0.74	0.921	0.890
8	WS, LCD,Lamp,Fan	0.88	0.731	0.911	0.870
9	WS, LCD,Laptop,Fan	0.839	0.669	0.892	0.862
10	**All 5 Appliances**	**0.768**	**0.614**	**0.879**	**0.842**

To evaluate the performance of our proposed models, the experiments are conducted in two phases: Binary and Multi-state operational phase. In the Binary phase, all the appliances are forced to operate only in their ON or OFF state disabling any intermediate state transitions (i.e., idle or standby). Conversely, for multi-state operations all possible state transitions have been taken into account for all target appliances. To train FHMM-AR and FF-AR models, separate data for each appliance has been collected during their binary and multi-state operations. A total of 10 test cases has been designed, each representing a unique set of appliance combination. An average duration of 30 minutes having approximately 120 events per test case have been generated by manually changing the respective states of the appliances. A 10 fold cross validation strategy has been adopted to report the performance of the models in terms of F-measure. In the subsequent section, we will summarize our findings from our experimental evaluations.

4.2. Results

The result for 10 different appliance combination scenarios has been summarized in Table (2). It has already been highlighted in our previous research work [9] that the performance of generative model FHMM-AR severely degrades under the following conditions

- Firstly, as the number of appliance states starts to grow the possibility of similar power draw values between appliance intermediate states also increases. This impacts the overall classification accuracy of the FHMM-AR models, as it can be seen from the results reported for test cases 5 to 10.

- Secondly, not only the number, but the type of appliances operating in parallel also has an impact performance of the FHMM-AR models. This can be seen from the results for all the test cases reported in Table (2), which contains a combination of fan, laptop and LCD. It is because FHMM-AR models makes an assumption during the model generation that the mean μ_t of respective appliance states linearly combine at the output as discussed in [9]. However, this is not true for cases when inductive and capacitive elements operate in parallel, such as fan and a laptop. Due to the leading and lagging power factor, the reactive powers of the load cancels each other instead of addition, thus violating the assumption. This is why, we see a low recognition performance of our generative model specifically for test cases $4, 8, 9$ and 10.

On the other hand, our discriminative model FF-AR perform significantly better than FHMM-AR models, as it can be seen from the results shown in Table (2). This is mainly because the classification accuracy of the discriminative models depends mainly on discriminative ability of the feature sets used to train the appliance models. FF-AR model is trained using a hybrid feature vector that increases the separability of appliance classes in the feature space, which consequently improves the AR accuracy. The keyboard typing sound of a laptop and the operational sound of a table fan, when jointly characterized with the power features using feature catenation helps removes the ambiguous overlapping between the two target classes. Therefore, we see a much higher F-measure for FF-AR model for test cases $4, 8, 9$ and 10 in comparison to FHMM-AR model as shown in Table (2). Although, we notice a performance degradation for the FF-AR model as the number of operating appliances increases, however if we look at test case 10 the FF-AR model achieves almost 11 % and 22 % improvement over FHMM-AR model for the case of binary and multi-state load classification, respectively. The overall average improvement achieved by the FF-AR model for all the test cases is about 3 % for binary and 10 % for multi-state device operations. This clearly indicates that unlike FHMM-AR models, the performance of the FF-AR models still maintain an acceptable performance even if the number of target appliances increases, however at the cost of increased complexity. We believe that although the use of additional audio modality increases the complexity of the system, however it is critical for the real-world systems to maintain an acceptable performance under all circumstances. Moreover, unlike other existing intrusive load monitoring systems reported in literature, our proposed multi-modal system as discussed in [10] maintains its non-intrusiveness by just employing only one audio sensor for the acoustic surveillance of target environment.

5. Conclusion and Future Work

This paper provides a comparison of generative and discriminative AR models in order to recognize the binary and multi-state operation of low-power appliances. We have demonstrated that our proposed discriminative model FF-AR is more suited for load recognition in comparison to our generative FHMM-AR model. Unlike FHMM-AR, the performance of FF-AR is less effected by the type and the number of target appliances operating simultaneously. The fusion of audio and power feature vectors clearly aids the classification process, resulting in a significant performance improvement for the FF-AR model. Our initial investigations demonstrate that in order to achieve acceptable AR performance in a real-world setting, it is promising to combine acoustic modality as a complimentary source of information, alongside power sensing. In future, we plan to extend this work by including more appliances in the dataset and further modify our algorithms for on-line load classification instead of performing off-line inference.

Acknowledgment
We acknowledge the support from the REDUCE project grant (no:EP/I000232/1) under the Digital Economy Programme run by Research Councils UK - a cross council initiative led by EPSRC and contributed to by AHRC, ESRC, and MRC.

References

[1] Uteley J and Shorrock L 2008 Domestic energy fact file 2008 Tech. rep. Building Research Establishment, Garston UK
[2] Ehrhardt-Martinez K, Donnelly K A and Laitner J A 2010 Advanced metering initiatives and residential feedback programs: A meta-review for household electricity-saving opportunities Tech. Rep. E105 American Council for an Energy-Efficient Economy (ACEE) Washington, DC
[3] Hart G W 1992 *In IEEE Proc.* **80** 1870–1891
[4] Zeifman M and Roth K 2011 *In IEEE Trans. Consum. Electron.* **57** 76–84
[5] Farinaccio L and Zmeureanu R 1999 *In Energ. Build.* **30** 245–259
[6] Liang J, Ng S K K, Kendall G and Cheng J W M 2010 *In IEEE Trans. Power Del.* **25** 551–560
[7] Suzuki K, Inagaki S, Suzuki T, Nakamura H and Ito K 2008 Nonintrusive appliance load monitoring based on integer programming *In Proceedings of SICE Annual Conference, Tokyo, Japan* vol 174 pp 2742–2747
[8] Zoha A, Gluhak A, Imran M A and Rajasegarar S 2012 *Sensors* **12** 16838–16866
[9] Zoha A, Gluhak A, Nati M and Imran M A 2013 Low-power appliance monitoring using factorial hidden markov models *In 2013 IEEE Eighth International Conference on Intelligent Sensors, Sensor Networks and Information Processing* pp 527–532
[10] Zoha A, Gluhak A, Nati M, Imran M and Rajasegarar S 2012 Acoustic and device feature fusion for load recognition *In Proceedings of 6th IEEE International Conference on Intelligent Systems (IS'12),Sofia, Bulgaria* vol 1 pp 386 –392

ICSICCST 2013 IOP Publishing
IOP Conf. Series: Materials Science and Engineering **51** (2013) 012003 doi:10.1088/1757-899X/51/1/012003

Participatory sensing as an enabler for self-organisation in future cellular networks

Muhammad Ali Imran[1,3], Ali Imran[2] and Oluwakayode Onireti[1]
[1] Centre for Communication Systems Research (CCSR), University of Surrey, GU2 7XH, UK
[2] Qatar Mobility Innovation Center, Doha, Qatar

E-mail: m.imran@surrey.ac.uk

Abstract. In this short review paper we summarise the emerging challenges in the field of participatory sensing for the self-organisation of the next generation of wireless cellular networks. We identify the potential of participatory sensing in enabling the self-organisation, deployment optimisation and radio resource management of wireless cellular networks. We also highlight how this approach can meet the future goals for the next generation of cellular system in terms of infrastructure sharing, management of multiple radio access techniques, flexible usage of spectrum and efficient management of very small data cells.

1. Introduction

Mobile cellular networks are traditionally designed to cater for the peak load/demand. This results in the wastage of valuable resources like energy, spectrum and space when the network is not fully loaded. This scenario is prevalent for a majority of the time when the network is active and hence the situation requires a more dynamic and intelligent design that will adopt the utilisation of the available resources according to the instantaneous load/demand of the system.

This challenge can only be met if we have a mechanism to monitor the system requirements and available resources in 'time' and 'space' domains in a live manner. On top of this we should have a mechanism to implement the on-the-run optimisation and configuration of network in an automated manner without the involvement of trained human resources to carry out the required tweaks and adjustments. The first requirement calls for a mechanism for collection of live data and the second requirement calls for the capability of self-organisation in the network.

Self organisation is a not a new area of research and development. The state of the art for this technique is discussed in detail in [1] where several defining features are also identified to classify a technique as self-organising. Some of the existing methodology towards self-organisation of cellular networks include: automatic neighbour relation function (ANR) for self-configuration of neighbour list in Long Term Evolution (LTE) system [2, 3]; self-configuration of propagation parameters such as antenna tilt and transmit power to improve coverage and capacity [4-6]; self-configuration of IP address [7, 8] and; load balancing through power and antenna adaption [9-11]. One of the key requirements of the self-organisation is to monitor and provide as an input all key parameters of the

[3] To whom any correspondence should be addressed.

system that affect the optimisation of configuration process. This can be achieved by enabling participatory sensing mechanism that can use the problem creators (the end user mobile devices and the applications running on them) to become part of a solution to the problem. Role of participatory sensing in a broad context is discussed in [12] and in the context of cellular/mobile devices is discussed in [13].

Participatory sensing offers several advantages over the conventional sensor networks which require the deployment of numerous static sensors, especially in urban areas. One main advantage of participatory sensing is that it leverages on the existing sensing provided by smart phones and communication provided by Wi-Fi and cellular architecture, thus, making the hardware deployment cost almost negligible. Also, participatory sensing inherits intrinsic economic scalability by exploiting existing phones as sensors. Furthermore, the open accessibility of application development toolkits for major mobile operating systems and convenient distribution channels in the form of App stores makes application development and deployment relatively easy. In addition, the inherent ubiquity and mobility of the phones yields spatiotemporal coverage at unprecedented scale which allows sensing and coverage of unpredictable events that might be missed by static sensor deployments [12]. The people-centred application of participatory sensing include: health and well-being where sensor enriched smart phones collect situ continuous data that improve the way wellness is reported and assessed as well as how treatment is delivered [14, 15]; monitoring sport experience; enhancing social media and price auditing [16-18]. Their environmental-centric applications include monitoring air quality, ambiance, noise, thermal column and road traffic condition [19-23].

In this short review paper we summarise the challenges that we anticipate in enabling this technique to be part of the future generation of cellular networks.

The rest of the paper is organised as follows. The next section presents the system model for participatory sensing. Section 3 summarised the challenges anticipated for making participatory sensing as a viable mechanism to gather data for enabling the self organisation of the network. Section 4 summarises the area of network management of optimisation that will benefit from the participatory sensing approach. Section 5 identifies the desirable goals and objectives for the design and implementation of the next generation of cellular networks and how they relate with the participatory sensing framework. Finally we summarise the paper and provide an outlook for future research and development in this area.

2. Participatory Sensing System Model

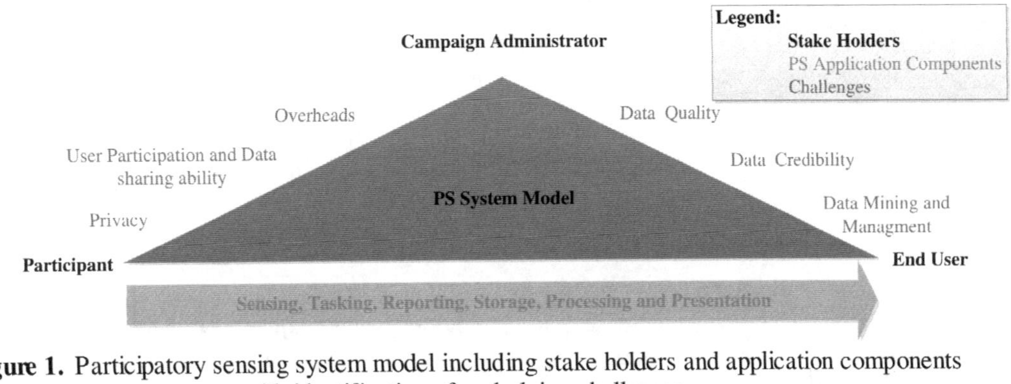

Figure 1. Participatory sensing system model including stake holders and application components with identification of underlying challenges

A generic system model of participatory sensing is presented in Figure 1. The system model is made up of three distinct stakeholders which are: the campaign administrators, the participants and the end users. The campaign administrators are responsible for setting up the participatory sensing campaign. They design, implement, and deploy the system architecture and are responsible for the maintenance and the management of the infrastructures. The participants have the sensing application on them and contribute to the participatory sensing campaign by gathering the sensor reading they have obtained. The end users on the other hand access and consult the data gathered by the participant according to their interests or preferences. End users could include the contributing participants which consult on their own collected data, campaign administrator verifying the obtained results or specialized entities attempting to gain insights about the monitored phenomena. Irrespective of the entity that serves as the end user, the reliability of the data produced by participatory sensing is a major concern on its application. The underlying challenges of participatory sensing can be categorised into two. The first category of challenges which include: user participation and sharing ability, privacy concern and overheads affects the main participatory sensing stakeholder, i.e. the participant. The second category of challenges primarily affects the administrator of a PS campaign, though it indirectly affects the 'end user' as well. This category includes challenges of controlling the spatiotemporal sampling to ensure quality of information, identifying and filtering out malicious or noisy inputs to ensure credibility of data gathered, and right context extraction without jeopardizing the participant. A detailed description of these challenges is presented in the next section.

3. Challenges in participatory sensing

Participatory sensing is a promising techniques but it is mired by several intertwined challenges that need to be overcome before it can provide its promised benefits, as shown in Figure 2. In the sequel we identify some of these main challenges and briefly discuss their scope.

3.1. Data Quality

Quality of the data collected by the participatory sensing techniques is of paramount importance to decide where and to what extent we can use this data for mission critical purposes. It is important to filter the data and identify the outliers as well as to fill in the gaps by interpolating and extrapolating the data to cover the missing parts of the full picture. These extrapolations also require methods for inference and analysis of the data. Finally all participating sensors do not have equal weightage in terms of their contribution so mechanisms need to be developed to apply user-preference and reliability measures while integrating the data from several sources.

3.2. Data Credibility

Participatory sensing relies heavily on altruistic user contributions. However, this openness makes participatory sensing vulnerable to malicious inputs. Without trust in the data provided by participants, the resulting statistics can be of little use. Therefore, it is important to ensure that the collected data comes from a trusted and credible source. Mechanisms and techniques need to be developed to ensure that any user added to the group or participating sensors should be admitted after authenticating their credibility. Mechanisms should also be in place to ensure that their performance and activity is monitored to ensure that untrustworthy adversaries do not hijack them.

3.3. Participant's Privacy

Participatory sensing enables availability of large amount of real time data shared by participants. This comes with an inherent challenge of maintaining watertight systems that can protect personal privacy. There is a need to develop a privacy-preserving framework that will enable users of participatory sensing applications to contribute data without having to worry about compromising their privacy. The research on privacy preservation in this context has to focus on both spatial and temporal privacy of users, which are two universal attributes that are expected to be included in user reports for all participatory sensing applications in general and participatory applications to be to be exploited for enabling self-organisation in particular. Another important aspect is to ensure the privacy of the participants' data while collecting the data in a live manner and also for the historic data stored in the cloud or other centralised devices.

3.4. User participation and data share-ability

From viewpoint of user participation and share-ability, participatory sensing can be considered analogous to a peer-to-peer network where users contribute sensing data to be shared among them or with central server. An issue that is central to both types of network is that there must be an incentive for the users to share, otherwise, it is well known from the "Tragedy of the Commons" that most users will tend to be free-riders rather than contributing resources to help system achieve its objectives. Therefore, one of the key challenges in participatory sensing is to convince and motivate the users to willingly participate in the data collection process. It is clear that this challenge is linked to the privacy challenge since ensuring the privacy of the end user's data can be a strong motivation for the users to be willing to participate in data sharing. Appropriate levels of abstraction and anonymisation should be implemented to convince users to share their personalised data (like user position, activity logs, trajectory and presence information in certain locations).

3.5. Data mining and management

The data collected by participatory sensing from a large scale of sensors and from a diverse group of participants is bound to scale very quickly into a large sized set – classified as a "big data" problem. Management of such a data set and extraction of useful information in a timely manner is a big challenge and an active area of research.

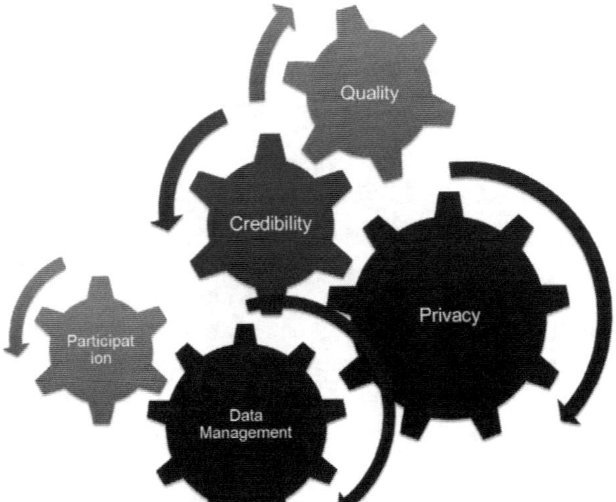

Figure 2. Different challenges in the implementation of participatory sensing actually drive and affect each other due to their intertwined nature.

A working participatory sensing mechanism at its own is not sufficient to provide full benefits for the optimisation of a multi-objective cellular network. This should be augmented with the mechanisms reliably communicating this information to a decision maker and providing means and mechanisms for data analysis and filtering. All these ingredients can ensure that we can optimise the system operation in light of multiple, and at times conflicting, objectives (like deployment optimisation, energy efficiency, spectrum efficiency and delay minimisation).

4. How participatory sensing will help?

Assuming that we have a working participatory sensing framework that collects the data from all users in a cellular system, we summarise different aspects of cellular network design and configuration that can benefit from this framework.

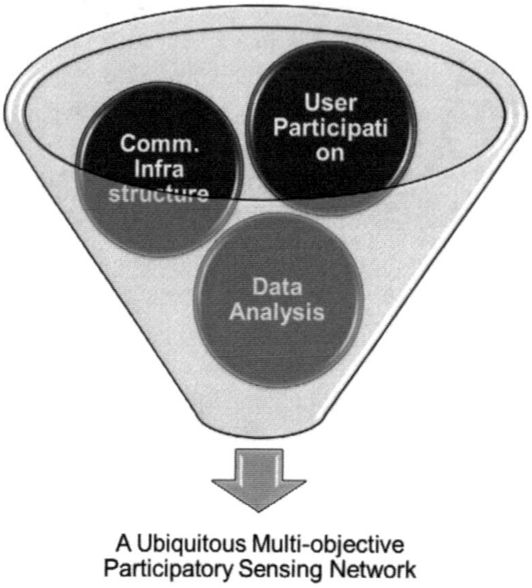

Figure 3. Design and implementation of the participatory sensing framework must have all the necessary ingredients to enable a ubiquitous multi-objective cellular network

4.1. Efficient Deployment

Data logged for user positioning and time stamps for data demand of different users at different times of the day can be utilised to identify the hotspots for service demand. In addition, signal levels along with the user positioning reported from the end user devices can also identify the gaps in coverage. All this information can benefit in optimising the deployment of infrastructure. This deployment can be a static deployment or can be changed in a more dynamic manner by switching ON/OFF certain nodes in the system depending on the demand and the time of the day as proposed in [24].

4.2. Efficient Self-organization

For a real-time, short time-scale optimisation and configuration of the network we require autonomous mechanisms for optimisation and configuration. This will also require real-time data acquisition from participatory sensing mechanisms. However, provided the data is reliable, credible and all users willingly participate, this can significantly improve the performance of the self-organising networks. An example of such self-organisation that can be based on participatory sensing data is autonomous antenna tilt optimisation while taking into account live user demography and activity profiles, as proposed in [25].

4.3. Efficient Radio Resource Management

Usually radio resources are managed on a long time-scale to adapt to average network and link statistics. However this may result in under-provisioning or over-provisioning of resources. However with a working participatory sensing mechanism in place, this can be drastically changed and the optimisation can track the variation in the opportunities very closely and hence reducing any associated wastage of the resources. Cell size or handover hysteresis optimisation for resource to load balancing based on live user experience is an example of possible application of such participatory sensing enabled efficient radio resource management [26].

4.4. Enabler for Future Architectures

Some futuristic architectures like massive MIMO, coordinated multipoint, opportunistic relaying etc. will require real-time channel state information and other system parameters to be reported to a decision making central point. All of these futuristic ideas will significantly benefit from the participatory framework for their self-configuration and optimisation.

5. Goals for future generation network

Based on the participatory sensing mechanism that will act as an enabler for self-organisation in future cellular systems, we should aim for several top level desirable features in the next generation of the system. Some of these desirable features are discussed briefly and it is highlighted how these ambitions will benefit from participatory sensing.

5.1. Infrastructure sharing

With the advent of very dense cellular systems for meeting the future data demands, we need to keep the spiralling capital expenditure CAPEX for each operator under control. For economic and technical viability of future cellular systems, the trend has to shift from competition to coopetition, among rivals in the industry. A simple model is to enable infrastructure sharing. A main impediment for enabling this is the perception of loss of control on the assets and inability of fairly sharing the revenues from the system. Participatory sensing can help by ensuring data collection regarding trajectories, usage patterns etc. for each operator's users to allow building up the profit sharing models as well as to allow a federation of assets where assets can be shared or kept for exclusive use based on the demands and dynamics of end users.

5.2. Co-existence of multiple radio access technologies

The network is fast becoming a melting pot of plethora of different options to provide wireless service. Despite of co-existing in space and time, so far these different options are operating mainly as competitors. Even if they are used to complement each other in some instances, the cooperation witnessed so far is ad-hoc and not well designed due to lack of availability of user specific data and inherent unpredictability associated with wireless eco-system. However, all this can change by exploiting the rich data collected from participatory sensing. Multi-vendor, multi-RAT co-existence exploitation agents can be designed to optimise the use of several radio access technologies jointly while taking into account user specific application requirements, social-economic portfolio and predictions of spatio-temporal footprints.

5.3. Flexible spectrum usage

One of the key features in future networks is more efficient and flexible usage of the spectrum. Some researchers propose the use of spectrum sensing based techniques for the flexible spectrum allocation but this will lead to unnecessary complexity and wastage of resources at the sensing end. On the other hand, some researchers are proponents of data base based spectrum allocation in time and space. However, creation of such databases is only feasible if we allow participatory sensing to collect and share the spectrum availability information that will again require spectrum sensing but this will only be needed during the stages when database is being populated. Once a database is stable, only few infrequent sensing operations will be required to keep the database up to date.

5.4. Management of individual data cells using control cells

An emerging architecture of dense cellular networks proposes the separation of data and control functionalities in the system. A large sized macro cell should control the coverage through control signalling and small, densely packed, access points should provide the data service. Such models are explained in details in [27, 28]. The working of this architecture can benefit from the participatory sensing by reporting the emerging demands and trends to optimise the hand-overs, data handling, switching ON/OFF certain data nodes and several other optimisation functionalities in future cellular networks.

6. Outlook and summary

Overall, we have identified that with emerging demands of very high efficiency in cellular networks the self organising mechanisms will become inevitable. One of the enabler for self organisation is timely availability of the optimisation/configuration parameters. In this regard we identify the potential role of participatory sensing techniques. We have discussed the intertwined challenges to enable participatory sensing framework and then identified the areas that can significantly benefit from such a mechanism. We have also identified some key features expected in next generation of cellular networks and discussed how participatory sensing can play a positive role in ensuring these features are implemented in next generation of the system.

7. References

[1] Aliu, O.G.; Imran, A.; Imran, M.A.; Evans, B., "A Survey of Self Organisation in Future Cellular Networks," *Communications Surveys & Tutorials, IEEE* , vol.15, no.1, pp.336-361, First Quarter 2013. doi: 10.1109/SURV.2012.021312.00116.

[2] Mueller C. M., Bakker H. and Ewe L., "Evaluation of the automatic neighbour relation function in a dense urban scenario," in Vehicular Technology Conference (VTC Spring), 2011 IEEE 73rd, 2011, 1-5.

[3] 3GPP TS 36.300, Evolved Universal Terrestrial Radio Access (E-UTRA) and Evolved Terrestrial Radio Access Network (E-UTRAN); overall description; Stage 2

[4] Wu J. S., Chung J. K. and Wen C. C., "Hot-spot traffic relief with a tilted antenna in CDMA cellular networks," IEEE Trans. Veh. Technol., vol. 47, no. 1, pp. 1 –9, Feb 1998.

[5] Temesvary A., "Self-configuration of antenna tilt and power for plug & play deployed cellular networks," in Proc. IEEE Wireless Communications and Networking Conf. WCNC 2009, 2009, pp. 1–6.

[6] Calcev G. and Dillon M., "Antenna tilt control in CDMA networks," in *2nd Annual International Wireless Internet Conference*, 2006.

[7] Alcatel-Lucent, "Automated installation of eNodeBs," R3-071251, 3GPP TSG RAN3, vol. Sophia Antipolis, 2007.

[8] Hu H., Zhang J., Zheng X., Yang Y., and Wu P., "Self-configuration and self-optimization for LTE networks," IEEE Commun. Mag., vol. 48, no. 2, pp. 94–100, 2010.

[9] Wang A. and Krishnamurthy V., "Mobility enhanced smart antenna adaptive sectoring for uplink capacity maximization in CDMA cellular network," IEE Trans. Commun., vol. 56, no. 5, pp. 743 –753, May 2008.

[10] Wu J., Bigham J., Jiang P., and Wang Y., "Intelligent control of cellular network coverage using semi-smart antennas for load balancing," International Conference on Signal Processing Communications and Networking, ICSCN '08, pp. 295 –301, 2008.

[11] Siomina I. and Yuan D., "Optimization of pilot power for load balancing in WCDMA networks," in IEEE Global Telecommunications Conference, 2004, (GLOBECOM '04), vol. 6, 2004, pp. 3872 – 3876.

[12] Burke J., Estrin D, Hansen M, Parker A, Ramanathan N, Reddy S, Srivastava M B, "Participatory sensing". In ACM Sensys World Sensor Web Workshop, Boulder, CO, USA, Oct 2006.

[13] Reddy, Sasank and Samanta, Vidyut and Burke, Jeff and Estrin, Deborah and Hansen, Mark and Srivastava, Mani, "MobiSense—mobile network services for coordinated Participatory Sensing", in International Symposium on Autonomous Decentralized Systems, 2009. ISADS'09. IEEE 2009.

[14] Blom J. and Hänninen R.. Air Pollution in Everyday Life: toward Design of Persuasive Urban Air Quality Services. 7th International Conference on Persuasive Technology (Persuasive), Linköping, Sweden. June 2012

[15] Consolvo S., McDonald D. W., Toscos T., Chen M., Froehlich J. E., Harrison B., Klasnja P., LaMarca A., LeGrand L., Libby R., Smith I. and Landay J. A., "Activity Sensing in the Wild: A Field Trial of UbiFit Garden", in Proceedings of the 2008 Conference on Human Factors in Computing Systems (CHI 2008), Florence, Italy, April 2008.

[16] Eisenman S. and Campbell A., "SkiScape Sensing", in Proceedings of the 4th International Conference on Embedded Networked Sensor Systems (ACM SenSys 2006), Boulder, Colorado, USA, November 2006.

[17] Gaonkar S., Li J., Choudhury R., Cox L. and Schmidt A., "Micro-Blog: Sharing and Querying Content Through Mobile Phones and Social Participation", in Proceedings of the 6th ACM International Conference on Mobile Systems, Applications, and Services (ACM MobiSys), pp. 174–186, Breckenridge, Colorado, USA, June 2006.

[18] Deng L. and Cox L. P., "LiveCompare: Grocery Bargain Hunting Through Participatory Sensing", in Proceedings of the 10th Workshop on Mobile Computing Systems and Applications (HotMobile 2009), Santa Cruz, California, USA, February 2009.

[19] Carrapetta J., Youdale N., Chow A. and Sivaraman V., "Haze Watch", online at http://www.pollution.ee.unsw.edu.au.

[20] Maisonneuve N., Stevens M., Niessen M. and Steels L., "NoiseTube: Measuring and Mapping Noise Pollution with Mobile Phones", in Proceedings of the 4th International Symposium on Information Technologies in Environmental Engineering (ITEE), Thessaloniki, Greece, pp. 215-228, May 2009.

[21] Bao X. and Choudhury R., "MoVi: Mobile Phone Based Video Highlights via Collaborative Sensing", in Proceedings of the 8th ACM International Conference on Mobile Systems, Applications and Services (ACM MobiSys 2010), Berkeley, California, USA, pp. 357-370, June 2010.

[22] Von Kaenel M., Sommer P., and Wattenhofer R., "Ikarus: Large-Scale Participatory Sensing at High Altitudes," in Proceedings of the 12th Workshop on Mobile Computing Systems and Applications (HotMobile), 2011, pp. 55–60

[23] Mohan P., Padmanabhan V., and Ramjee R., "Nericell: Rich Monitoring of Road and Traffic Conditions using Mobile Smartphones," in Proceedings of the 6th ACM Conference on Embedded Network Sensor Systems (SenSys), 2008, pp. 323–336.

[24] Capone, Antonio and Filippini, Ilario and Gloss, Bernd and Barth, Ulrich, "Rethinking cellular system architecture for breaking current energy efficiency limits", in Sustainable Internet and ICT for Sustainability (SustainIT), 2012, IEEE.

[25] Chen, Yan and Blume, Oliver and Gati, Azeddine and Capone, Antonio and Wu, Chi-En and Barth, Ulrich and Marzetta, Tom and Zhang, Haibin and Xu, Shugong, "Energy Saving: Scaling Network Energy Efficiency Faster than Traffic Growth" Wireless Communications and Networking Conference Workshops (WCNCW), 2013.

[26] Imran, A.; Imran, M.A.; Tafazolli, R; "A novel Self Organizing framework for adaptive Frequency Reuse and Deployment in future cellular networks," 21st Radio Communications (PIMRC), IEEE; vol., no., pp.2354-2359, 26-30 Sept. 2010

[27] Imran, A.; Imran, M.A.; , Atta-ul-Quddus; Tafazolli, R.; , "Distributed spectral efficiency optimization at hotspots through self-organisation of BS tilts," IEEE GLOBECOM, 2011, vol., no., pp.570-574, 5-9 Dec. 2011

[28] Imran, A.; Yaacoub, E.; Imran, M.A.; Tafazolli, R.; , "Distributed Load Balancing through Self Organisation of cell size in cellular systems," Personal Indoor and Mobile Radio Communications (PIMRC), 2012 IEEE 23rd International Symposium on , vol., no., pp.1114-1119, 9-12 Sept. 2012

Acknowledgments

This work was made possible by NPRP grant No. 5 - 1047 - 2 – 437 from the Qatar National Research Fund (a member of The Qatar Foundation). The statements made herein are solely the responsibility of the authors.

Fabrication and Structural Characterization of Co-implanted Ultra Shallow Junctions for Integration in Piezoresistive Silicon Sensors Compatible with CMOS Processing

S Ahmed[1,2] and R Mustafa[1,3]
[1]Advanced Electronics Laboratory, Faculty of Engineering & Technology, International Islamic University, Islamabad - 44000, Pakistan.
[2] Centre for Emerging Sciences, Engineering & Technology, Islamabad-44000, Pakistan.
[3] Department of Physics, International Islamic University, Islamabad-44000, Pakistan.

E-mail: ahmed.shuja@iiu.edu.pk

Abstract. Fabrication and structural characterization of Indium and Carbon implanted n-type Silicon layers forming ultra-shallow junction for integration in piezoresistive sensors compatible with CMOS processing is studied in detail. The co-implantation technology together with medium range annealing temperature regimes seem to play an important role at atomistic level and provide a process control to engineer the strain and maintain the quality of surface/layer/active device region for further manufacturing process cycle. This is likely to impact the yield and reliability for the fabrication of these devices for diverse applications.

1. Introduction

Piezoresistors are commonly used in accelerometers, pressure sensors and different kinds of sensor arrays. These devices have also emerged as the industry's choice while manufacturing the Scanning Probe Microscopy (SPM) and Atomic Force Microscopy (AFM) systems. The manufacturing industry of these systems utilizes the piezoresistors to sense the deflection in cantilevers [1-4]. A commercial piezoresistor consists of doped (diffused or ion-implanted) silicon which in comparison to rest of the materials exhibits remarkable and optimized response characteristics [5-6]. Realization of large area and fast high resolution imaging is incomplete without a sensible integration between various components of microsystem (piezoresistive sensors, actuators, CMOS processing compatible cantilever, etc.). The creation of various doped layers, regions with tailored physical properties, engineered surfaces, stable interfaces and ultra-shallow junctions are extremely important to realize such Microsystems, particularly the ultra-thin and fully integrated piezoresistive sensors.

It is important to ascertain the maintainability of the material properties of the junction after a certain process recipe employed in fabricating the Ultra Shallow Junctions for CMOS nano-electronics [7, 8, 9]. The processing protocol to examine the formation and subsequent electrical characterization of under study co-implanted Si layers (doping concentration, sheet resistance, mobility and junction depth) is already reported in depth [9]. In order to fully optimize the test process and fabrication strategy, a controlled behavior of tailored films and surfaces on atomistic level is required to ascertain process yield and manufacturing cost. X-rays (in various forms such as XRD, XRF, X-rays scattering measurements etc.) are widely used in semiconductor manufacturing industry to examine, evaluate, and analyze the surfaces, interfaces, thin films and device regions. XRD, in particular, has emerged as standard diagnostic technique for structural investigation in ion implantation technology particularly to study the stress, strain and distortion created in the ion implanted layers due to the creation of point defects and/or extendd defects in the crystal. Some End-of-Range (EOR) defects may reportedly be responsible for the clustering of silicon interstitials in this region where indium dopant profile dominates in the Si matrix [7, 10, 11]. In the similar fashion, diffusion of carbon into silicon has also been reported to act as an effective sink for silicon interstitials [12, 13, 14, and 15]. XRD studies of Si-layers with Indium and Carbon co-implantation with pre-defined post-implant annealing schedules were also carried out with a reported recovery of Si lattice constant to its original value by post-implant annealing cycles (600°C-1050°C) [10]. Further, the reduction in strain with increasing indium dose corresponding to a reduction in the dopant deactivation was also reported in detail [16]. The aim of this paper is to present the findings on the reliability (in terms of quality of the processed material for direct utility in Si-based piezoresistors and consequent integration with rest of the processes to achieve a total device solution) using structural techniques with special reference to the surface roughness and structural morphology. Such strain engineering in nanoscale materials, surfaces and devices as well as quantum structures is applied as an important and powerful process strategy to develop improved functionalities and high performance devices [17, 18, and 19].

2. Experimental Conditions, Design and Details

Detailed experimental design is elaborated in one of the previously published work [9]. The important experimental parameters are tabulated below (Table 1). The implant parameters in this work (Also; previously used to form junctions reported in references 8 and 9) are purposely chosen to be similar or close to those studied earlier [7 and their relevant work] in order to check the reliability of the physical process as well utilize their complimentary analysis to improve the process's functionality requirements.

Table 1: Co-implantation Assisted USJ Formation in Silicon: Design & Fabrication Parameters

Process/Fabrication Step	Parameters
Selection/Growth of Starting Material	Commercially grown, CZ n-type Si wafers of (100) orientation with initial resistivity between, 4-7 Ω/square
Activation-specific Pre-amrphization Implantation	Si at an energy of ~500eV and a fluence of ~5×10^{15} at/cm^2
Junction Formation: Ion Implantation	Ion: Indium (In+); Dose: 5.7×10^{14} at/cm^2; Implant Angles: 8° tilt and 24° twist; Energy: 70 keV
Junction Formation: Co-Implantation	Ion: Carbon (C+); Dose: 3.4×10^{15} at/cm^2; Implant Angles: 8° tilt and 24° twist; Energy: 10keV
Cleaving	squares of 1-1.5cm^2 sizes
Dopant Activation via Rapid Thermal Processing	Rapid Thermal Annealing (RTA) between 600 and 800°C for 60s dwell and carefully designed ramp-up time

Atomic Force Microscopy and XRD facilities were used to image the extent of surface roughness (structural morphology) and scan the diffraction data revealing the structural changes due to ion irradiation of the as-implanted and annealed Si layers. The machine parameters are summarized in Table 2.

Table 2: Process Control and Precision Check by Structural and Imaging Techniques: Parameters

Technique	**Measurement Parameters**
X-ray Diffraction	Rotating anode X-ray diffractometer (Philips X'pert); CuKα radiation of wavelength 0.154 nm
Atomic Force Microscopy	AFM Agilent system 2500 picoscan; Maximum Scan Rate: 48 lines/sec at 256x256; 100*um* with large scanner and atomic resolution with small scanner

3. Results, Analysis and Discussion

The 2θ diffraction scans of as-implanted samples of co-implanted Si layers were taken to identify the structure and phase. The assignment of signal peaks at ~33° and 39° was made to Indium (111) and (110) diffractions, respectively. Similarly; the peaks observed at ~43° and 44.5° were assigned to Carbon (100) and (101) respectively. The XRD spectra shown in figure 1 exhibits the presence of C, In and Si atoms in n-type Si matrix. While looking at the spectra carefully, one notices that the as-implanted samples (un-annealed) have relatively a maximum intensity at ~69.13° assigned to Si (400). Further, the effect of post-implant annealing on co-implanted layers is obvious when samples are annealed at T_{anneal}= 600°C. A sharp peak appears as a little hump with a change in the extent of relative intensity of In (110). This may be a signature of the physical process being activated on atomistic level suggesting an increase electrical activation efficiency of indium atoms at 600°C, hence a lowering intensity of Si (400) as compared to un-annealed cases. We know from a previously reported study [2] that the junction depth increases in the devices under discussion in the temperature ranging from 600°C to 700°C. This increase in junction depth may be attributed to the movement of indium atoms back to their original positions [20]. Interestingly, a slightly different behaviour of physical processing at atomistic level is observed when we anneal the samples further at 750°C. The signature of this process emerges in the form of increase in lattice strain and hence the junction depth decreases [20]. The breaking of C-In complexes may be responsible for this structural activity. At 800 °C, the ion-implantation-induced strain in the lattice seems to decrease. Ion related strain is known to be removed almost fully by high temperature annealing in Si. The implantation-induced strain in the lattice and activated indium atoms present near the substitutional lattice sites (and hence forming the C-In acceptor centers) and their subsequent physical behavior while undergoing the post-implant annealing schedules play an important role in making the doped Si layers either relaxed or strained directly impacting the fabrication yield and efficiency of piezoresistors integrated with the Microsystems. There are certain evidences that the vacancies also interact with substitutional Indium atoms to relax the strain field duly introduced by the relatively larger covalent radius of In ~0.144 nm compared to that of Si ~0.11 nm [20]. Also, it has also been observed that the dopants (indium and carbon in this case) in Si may be responsible to make the lattice contracted or expanded depending on the host-impurity covalent bond radii [16]. Our results summarized in figure 2 also augment these findings where strain caused due to co-implantation in Si layers vary with increasing annealing temperature and consequently impact the depth of the junction formed.

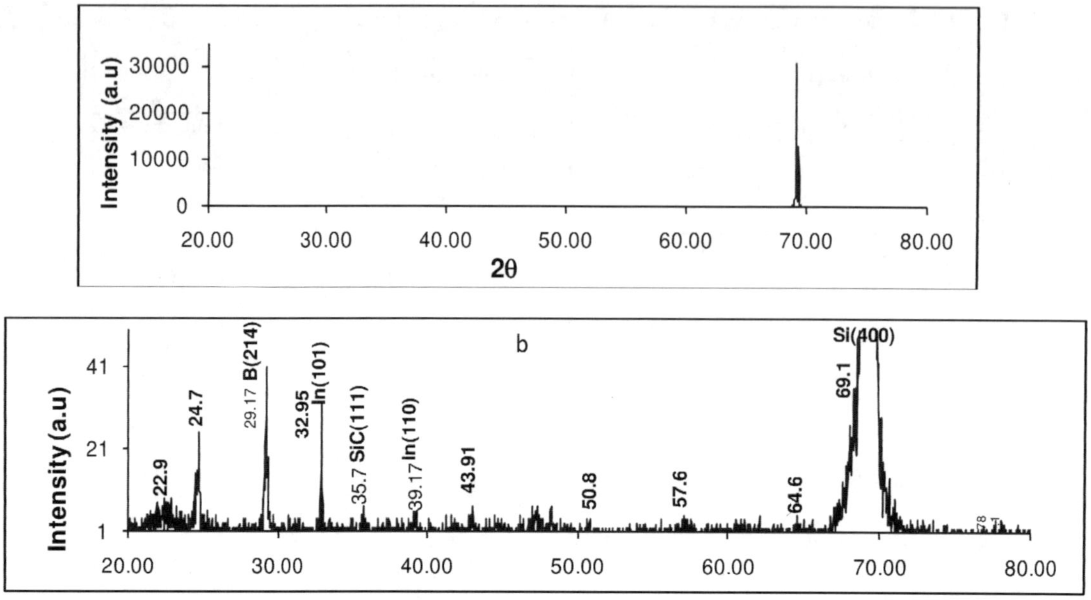

Figure 1. XRD Spectrum of In and C co-implanted Si layers undergone a rapid thermal annealing process at temperature regime 600-800°C. (a) as-implanted (b) annealed at 600°C (c) annealed at 650°C (Spectra obtained for other annealing temperatures are not shown here due to space limitations. The key features/trend, however, are summarized in figure 4)

Surface morphology of a material/layer/thin film is an important structural parameter to assess the quality and utility of any fabrication process undergoing a full-device cycle. At present, the most promising approach to study an in-depth morphology is Atomic Force Microscopy. Surface representing a non-Gaussian distribution and an exponential autocorrelation function may be examined by Atomic Force Microscopy. An output measurement parameter RMS (the root mean square of the vertical deviations from the reference) is connected with the distribution's probability density function, and serves as a quick evaluation tool to analyze the pre-and post-processed surface morphologies [21].

RMS roughness tends to increase with the increasing post-implant annealing temperature as exhibited in figure 3. Figure 3 demonstrates the topographical view of the In and C co-implanted Si layers before annealing reflecting the extent of roughness and uniformity of the irradiated surface. The presence of impurity concentration at different depths is also visible as background structural information. This three dimensional view shows negligible degradation of the morphology of the surface after the devised and processed implant strategy. Positioning of thermal flux due to annealing considerably reduces damages (refer to Figure 4) and is crucial to study the consequent damage re-engineering in Si-based devices, particularly in thin/ultra-thin doped/damaged layers (utilized in the fabrication of piezoresistors), shallow/ultra-shallow junctions (utilized in source or drain engineering for Si-CMOS manufacturing as well as integration with piezoresistors based Microsystems) and region definition for contact fabrication. Figure 4 shows the variation of RMS as a function of annealing temperatures for ion implanted Si layers. The values reported do averages of six different measurements at different locations (regions of the surface) of the sample, clearly reflecting a trend owe the changes on the atomistic level analogous to the finding and discussion on XRD measurements.

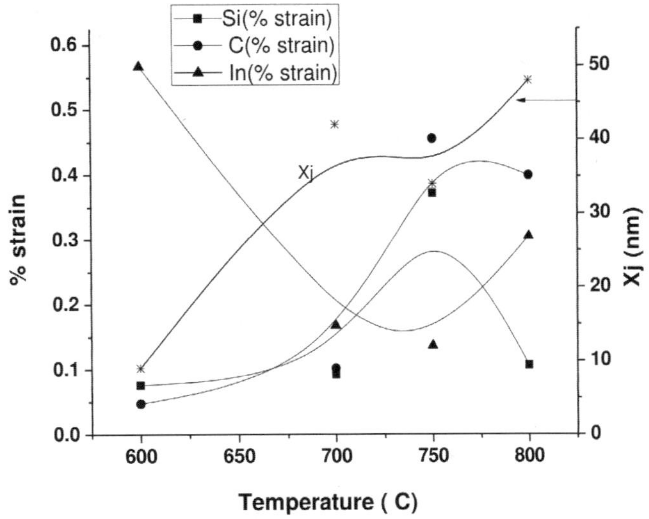

Figure 2. Variation of % change in Strain obtained from XRD as a result of In and C co-implantation in n-type Si layer with respect to annealing temperature ranging from 600°C to 800°C. Junction depth obtained on these samples by various complimentary techniques reported previously [8,9] are also plotted to correlate the results.

Figure 3. Representative Atomic Force Microscopic (AFM) topographical view of the In and C co-implanted Si layers (un-annealed): Extent of roughness and uniformity of the implanted indium and carbon co-implants. This three dimensional viewgraph shows negligible degradation of the morphology of the surface after the devised and processed implant strategy for effective utilization in the formation and integration of piezoresistive sensors and CMOS ultra-shallow junction.

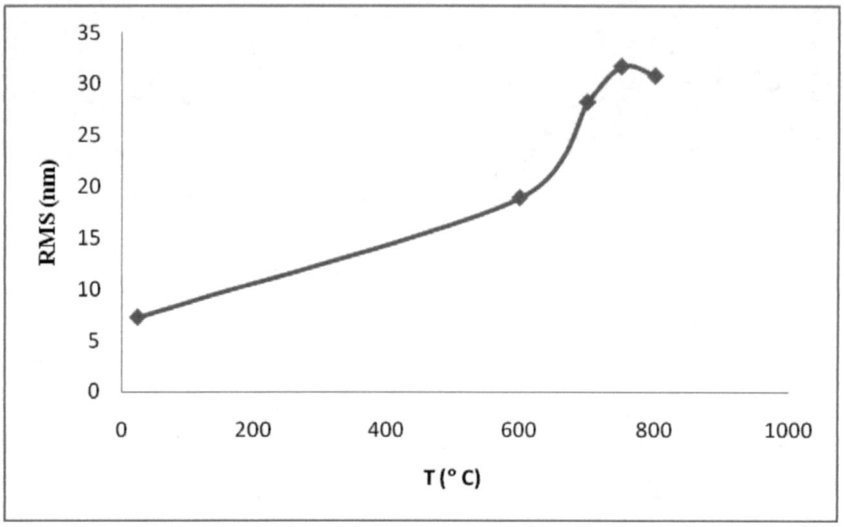

Figure 4. Variation of surface roughness amplitude parameter as a function of post-implant annealing temperature for co-implanted silicon layers

4. Conclusion

In summary, we studied the thermal effects on structural and morphological properties of the layers engineered by co-implantation technology in Si to provide a controlled, reliable, repeatable and easy-to-drive process to fabricate ultra-thin piezoresistors and ultra-shallow junctions for CMOS compatible fully integrated systems. This study suggests that a careful design of implant-annealing processes may provide a useful process control directly impacting the physical activity at atomistic level such as defect clustering, defect migration and annihilation, movement of dopant atoms within the atom distribution profiles, and surface roughness, lattice relaxation and strain duly controllable by the residual thermal flux during the annealing process.

Acknowledgement

The author would like to acknowledge the concerned responsible users and principal investigators of XRD, AFM and other complimentary facilities at the COMSATS Institute of Information Technology, Islamabad, Pakistan, Quaid-i-Azam University, Islamabad, Pakistan, Berkeley National Laboratory, USA and Michigan State University USA for repeated measurements (some of the measurements were also cross-checked on machines other than the reported ones). Core Systems and IBS, USA area also acknowledged providing necessary support to carry out the implantation schedules. This research was supported by HEC, Pakistan under indigenous scholarship program as well as NRPU's funding for USJ project.

References

[1] Picco L M, Bozec L, Ulcinas A, Engledew D J, Antognozzi M, Horton M A, and Miles M J 2007 *Nanotechnology* **18** 044030

[2] Minne S C, Yaralioglu G, Manalis S R, Adams J D, Zesch J, Atalar A, and Quate C F 1998 *Appl. Phys. Lett.* **72** 2340

[3] Vettiger P, Cross G, Despont M, Drechsler U, Durig U, Gotsmann B, Haberle W, Lantz M A, Rothuizen H E, Stutz R, and Binnig G K, 2002 *IEEE Trans. Nanotechnol.* **1** 39

[4] Salaita K, Wang Y H, Fragala J, Vega R A, Liu C, and Mirkin C A 2006 *Angew. Chem.* **45** 7220

[5] Smith C S 1954 *Physics Review* **94** 42

[6] Kanda Y 1991 *Sensors and Actuators A: Physical* **28** 83
[7] Gennaro S, Sealy B J and Gwilliam R 2005 *Electron. Lett.* **41** 1302
[8] Rehana Mustafa et al 2013 *J. Basic Appl. Sci. Res* **6** 178
[9] Rehana Mustafa et al 2013 *Chin. Phys. Lett.* **30** 016101-1
[10] Foong C T et al 2003 *J. Appl. Phys. Lett.* ***83*** 4169
[11] Noda T 2002 *J. Appl. Phys.* ***91*** 639
[12] Rucker H et al 1998 *J. Appl. Phys. Lett.* ***73*** 1682
[13] Rucker H et al 1999 *J. Appl. Phys. Lett.* ***74*** 3377
[14] Carrol M S and Sturm J C 2002 *J. Appl. Phys. Lett.* ***81*** 1225
[15] Shishkin M et al 2004 *Mat. Res. Soc. S.* ***810*** C8 4 1
[16] Clarka H M and Jones S K 2005 *J. Appl. Phys.* ***97*** 093525
[17] Chu M et al 2009 *Annu. Rev. Mater. Res.* ***39*** 203
[18] Zhigang W et al 2009 *Nano Lett.* ***9*** 2418
[19] Moutanabbir O et al 2010 *Phys. Rev. Lett.* ***105*** 026101
[20] Leea M L and Eugene A 2005 *J. Appl. Phys. Rev.* ***97*** 011101
[21] Chandrasekaran S and Sundrarajan S 2004 *J. Surf. Coat. Tech* **188** 581

IOP Conf. Series: Materials Science and Engineering **51** (2013) 012005 doi:10.1088/1757-899X/51/1/012005

Process Simulation and Characterization of Substrate Engineered Silicon Thin Film Transistor for Display Sensors and Large Area Electronics

S M Hashmi[1,2] **and S Ahmed**[1,2]
[1]Advanced Electronics Laboratory, Faculty of Engineering & Technology, International Islamic University, Islamabad - 44000, Pakistan.
[2] Centre for Emerging Sciences, Engineering & Technology, Islamabad-44000, Pakistan.

E-mail: ahmed.shuja@iiu.edu.pk

Abstract. Design, simulation, fabrication and post-process qualification of substrate-engineered Thin Film Transistors (TFTs) are carried out to suggest an alternate manufacturing process step focused on display sensors and large area electronics applications. Damage created by ion implantation of Helium and Silicon ions into single-crystalline n-type silicon substrate provides an alternate route to create an amorphized region responsible for the fabrication of TFT structures with controllable and application-specific output parameters. The post-process qualification of starting material and full-cycle devices using Rutherford Backscattering Spectrometry (RBS) and Proton or Particle induced X-ray Emission (PIXE) techniques also provide an insight to optimize the process protocols as well as their applicability in the manufacturing cycle.

1. Introduction

Ion beam technique is widely used as a flexible and powerful tool for material's defect and substrate engineering [1], and Silicon ICs and transistor fabrication. The controlled introduction of implantation-induced defects is also used to tailor specific electrical and optical properties of the material and devices [2, 3]. Top gate Thin Film Transistors (TFTs) are generally manufactured by exploiting the physical characteristics of amorphous silicon films. The commercial TFTs used in Liquid Crystal Display (LCD) and Large Area Electronics (LAE) technologies have staggered bottom gate geometry with the channel passivated by SiN_x [4]. The physical platforms (both as-grown and engineered using ion implantation technology) of the transistors are usually annealed after fabrication for device and electrical efficiency activation [4]. Fabrication of such devices is made possible with many process recipes including the coplanar amorphous-Si:H TFT on glass [5] and inverted staggered TFT with a combination of SiN_x , *a*-Si:H and n^+a-Si:H [6].

2. Si substrate engineered TFT: Model and Design

The TFT device design was modeled using commercially licensed SILVACO TCAD software. An inverse staggered bottom gate *a*-Si TFT on 200nm thick Si substrate was used as test structure to imitate a commercial device. First a layer of amorphous silicon (*a*-Si) followed by a 300nm Si_3N_4 layer, and 200 nm thick silicon dioxide layer, were sequentially deposited at 300^0C by Plasma Enhanced Chemical Vapour Deposition (PECVD) [7-8]. Then, source and drain regions are defined using aluminum by photolithography and the oxide is etched off from these regions. Implantation of 130 keV Helium (He^+) and 160 keV Silicon (Si^+) ions with doses from $1\times10^{13}cm^{-2}$ to $1\times10^{16}cm^{-2}$ and $1.5\times10^{15}cm^{-2}$, was carried out at commercial implanter, respectively. During the implantation experiments, a tilt of 7^0, rotation of 22 to 24 degrees and nominal room temperature was purposely maintained. The key design parameters are summarized in Table 1, whereas the corrected, simulated and modeled TFT designs are presented in Figures 1 and 2 below.

Table 1: Design Parameters of TFT Fabrication

Process Steps	TFT Design parameters
Platform/Substrate: Cz-Silicon	<100> orientation Depth 0.2um (200nm) With doping of phosphorous with sheet concentration=$1.0\times10^{13}cm^{-2}$
Oxide Deposition: oxide layer(SiO_2)	0.20um (200nm)
Etching: oxide layer(SiO_2)	left p1.x=2.0
Nitride Deposition: (Si_3N_4)	0.30um (300nm)
Amorphous Silicon Definition	0.20um (200nm)
Substrate Engineering Implants: Implant He^+ and Si^+ ions	Helium ion doses=(1×10^{13}, 1×10^{14}, 1×10^{15}, 1×10^{16}) cm^{-2} energy=130keV, Silicon dose=1.5×10^{15} cm^{-2} energy=160keV crystal tilt=7° rotation=24°
Source/drain contact Metallization: Deposition of Aluminum (Al)	thickness=0.2um
Etching of Aluminum (Al)	left p1.x=0.8 right p1.x=2.4
Mirror to get full structure	Structure defined: mirror right
Electrodes specifications	Source x=1.0 y=-0.6 Drain x=5.0 y=-0.6

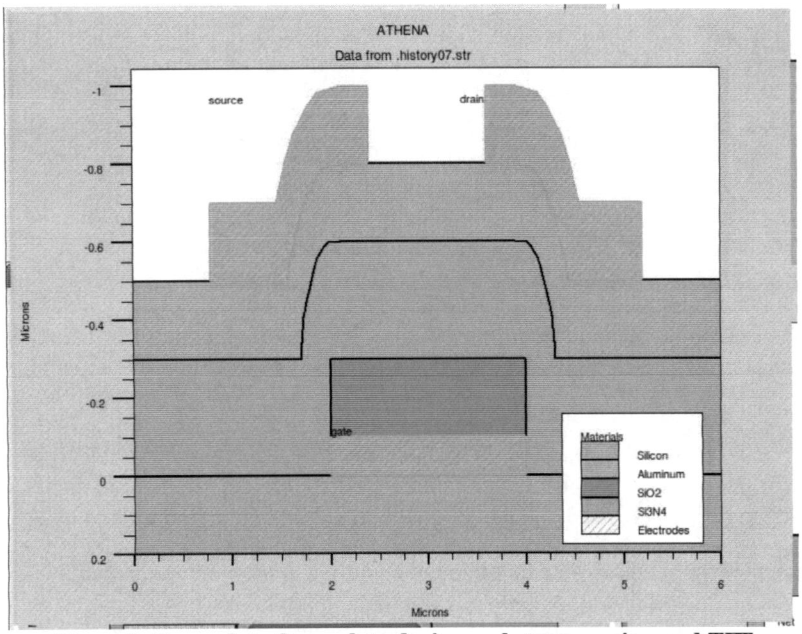

Figure1. Tony plot of complete device: substrate engineered TFT

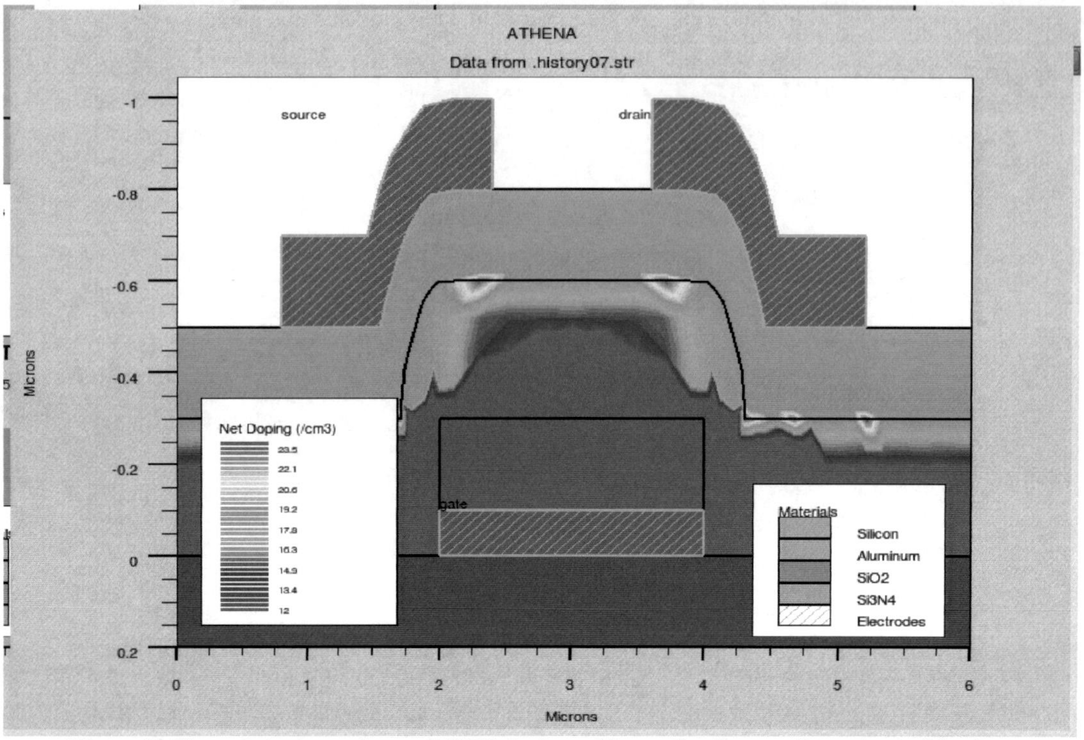

Figure2. The substrate engineered TFT design with a reflection on the extent of doping

3. Post-process qualification of substrate engineered devices

Rutherford Backscattering (RBS) [9] and Particle/Proton Induced X-ray Emission (PIXE) [10] techniques are used for post-process qualification of devices and engineered material for possible LCD sensing and LAE applications. This is important for the qualification of our proposed design in order to address the pertaining issues in fabrication and full-cycle manufacturing of the device. Both RBS and PIXE measurements are performed in Accelerator Laboratory, National Centre for Physics, Islamabad. The general parameters used for Rutherford backscattering are shown in table 2. Standard RUMP software is used to analyze the RBS data presented in Figures 3 and 4. RBS spectra are obtained by bombarding $^4He^{++}$ ions with energy of 3MeV and scattering angle of 170° on N-type (Phosphorous doped; refer to Table 1) Cz-Si wafers. In the helium and silicon implanted Si wafers, no peaks are detected up to 3.5um, because helium cannot be detected with conventional RBS technique. Less than 1% concentration of phosphorus peak due to lightly doped N-type Si wafer is observed in Fig. 3 whereas the post-annealed (~900-1000°C regime) full-cycle TFT device reveals a further loss of phosphorous (owing to other dopants such as Galium to define p-type regions) to 0.5% in Fig. 4. Approximately equal amount of backscattering yield shows Gaussian distribution in device suggesting a uniform damage distribution across the active device or a pseudo-amorphized region. The calculated Kinematic factor is 0.77. The highest normalized yield is obtained at 0.31MeV at channel number 256 (refer to Fig. 3). The percentage composition of silicon is 99.5% and phosphorous is only 0.5 to 1%%. The analyzing depth is 3500nm.

In the present work, PIXE analysis has been used to analyze the presence of in-situ or post-process impurities/contaminations in the device, known to affect the efficiency of electrical characteristics of the TFT and response time in display sensors. The incident beam of protons (H^{+1}) of typically of

2.5MeV energy (beam size: 2mm, beam current: 3nA and charge: 5μC) is used for PIXE analysis. Si (Li) x-ray detector is used with energy resolution 180eV and detector to sample distance is about 12.5cm. Figure 5 reveals a very less amount of contaminations of Ca, Ti and Zr found in all the analyzed samples. The backscattering yield of Calcium varies from 40 to 290, Titanium ranging from 140 to 800 and Zirconium from 20 to 200. The intensity of these contaminations is very low and negligible to affect the overall device characteristics in any significant ways, as depicted in Figures 5 and 6.

Table 2: RBS Parameters used in post-qualification experiments

Ion Beam	He^{++}
Beam Energy	3MeV
Beam Size	2.00mm
Analyzable Elements	Be^{+} to U^{+}
Analyzing Depth	3*u*m
Charge	20*u*C
Beam Current	≈55nA
Scattering Angle	170°
Incident Angle	7°
FWHM	15.0KeV
Sensitivity	10 parts per million

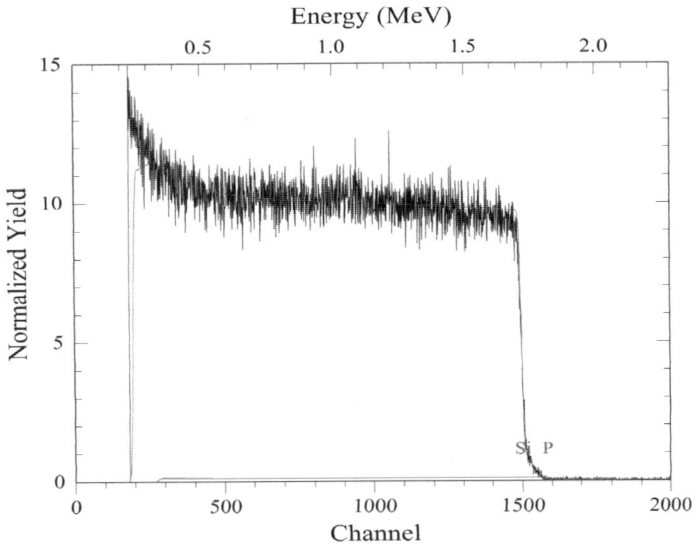

Figure 3. RBS spectrum: Substrate Engineered Silicon (Starting Material with the damage implants with parameters given in Table 1)

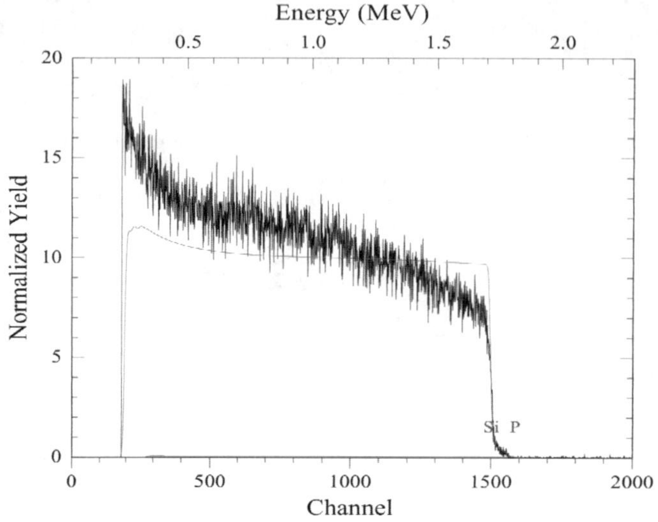

Figure 4. **RBS spectrum: Substrate Engineered Silicon TFT-Full cycle post-process annealed device (Starting Material with the damage implants with parameters given in Table 1)**

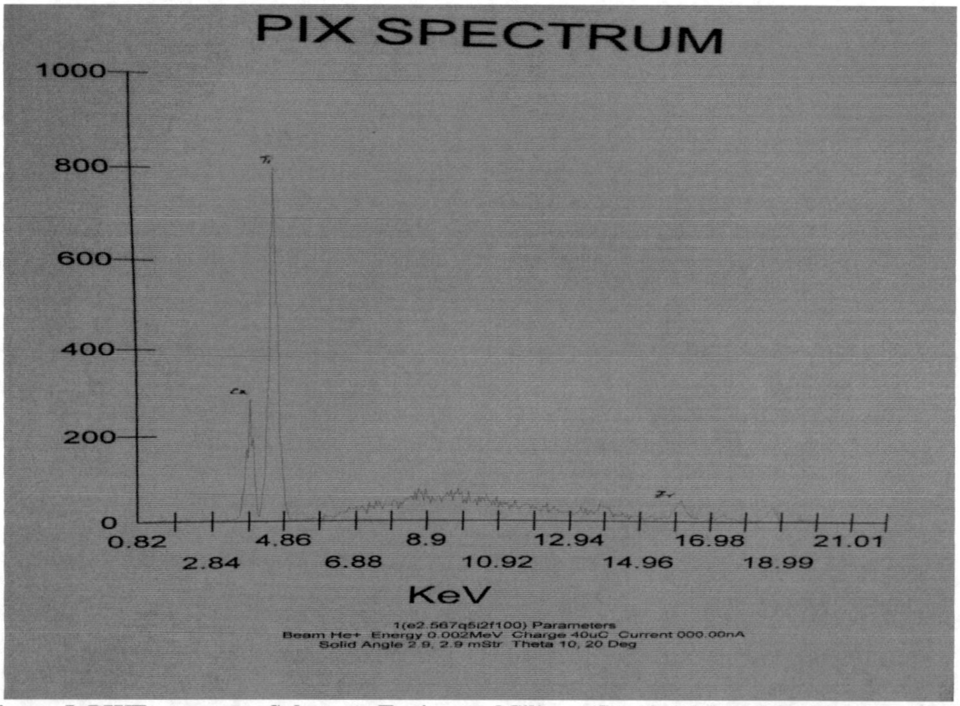

Figure 5. **PIXE spectrum: Substrate Engineered Silicon (Starting Material with the damage implants with parameters given in Table 1)**

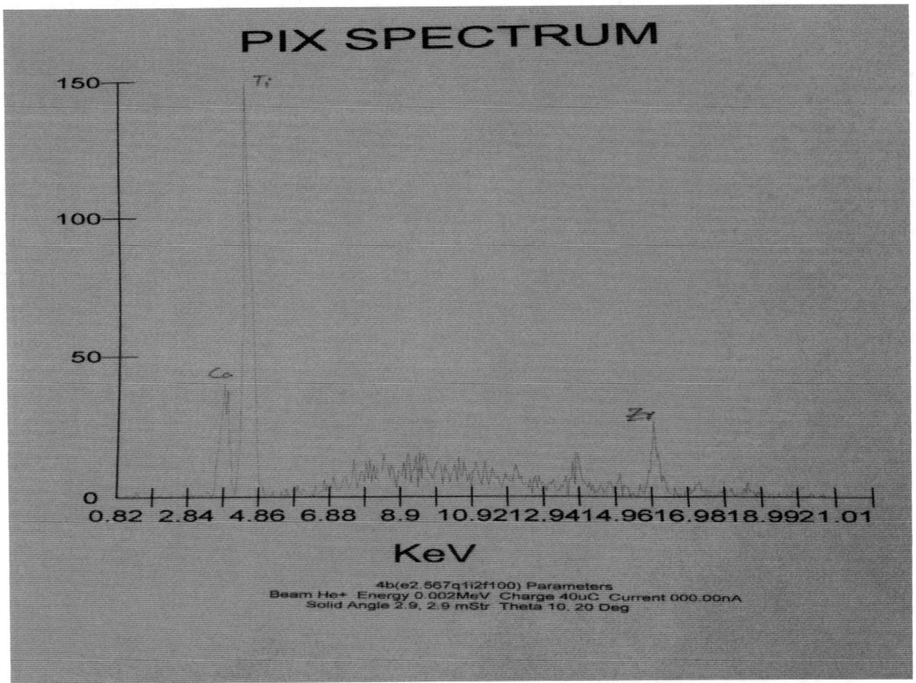

Figure 6. PIXE spectrum: Substrate Engineered Silicon TFT-Full cycle post-process annealed device (Starting Material with the damage implants with parameters given in Table 1)

4. Results and Discussion

While doing RBS measurements, it provides us a chance to analyze the collision kinematics, the energy loss and analytical Rutherford cross-section. The collision kinematics, electronic part of the energy loss of the helium beam in the Si-matrix and the yield (connected to the analytical Rutherford Backscattering) refer to the mass of the constituents in the matrix (Si and P in our case), where (how deep and in what form) they are found in the substrate and how much (The content percentage), respectively.

The spectral variation in all such cases, where implantation of He^+ and Si^+ into n-type Si-wafer is carried out, is minimal up to a certain depth resolution (surface and near surface region of interest), This is valid owing the fact that the damage created due to helium implants together with the consequent damage of the lattice due to silicon implantation acted in such a balance that the whole region of interest behaved as a flat/nearby uniformed damage distribution causing a pseudo-amorphous layer exhibiting the properties of amorphous silicon material generally used for LCD/TFT environments. This is indeed a very important physical parameter while optimizing the response time in the display sensors based on TFT/LCD technologies.

The similarity of spectral variation amongst the substrate engineered devices represented reveal that the helium or silicon atoms have sufficiently created a damage which is solely responsible for the subsequent creation of the amorphized region in the active device area. The RBS resolution and channel information prove this understanding for all such cases without a single trace where the displacement/replacement caused a localized trace of evolution of crystalline region or boundary. In Fig. 4: the post process and annealed devices show a loss of phosphorous by 0.5% for relatively high thermal treatment and a slightest recovery of the amorphization region. No substantial variation is exhibited for annealing performed at lower temperatures such as $T_{anneal}=500^0C$ (not shown here); which in silicon devices does not provide a defect-annihilation process window.

The combination of He^+ and Si^+ damage implants seems to be perfect way to hinder Si^+ from being dopant into Si-substrate either on" as-grown damage/defect sites" or because of deficient solubility in the matrix. He^+ is a perfect choice for attaining the right kind of damage extent to create a uniform damage profile necessary for an amorphous layer. This "engineered amorphized region" is created due to carefully designed implantation experiment by varying process parameters such as energy, selection of ion specie, fluence etc. Moreover; the extent and nature of damage caused due to physical bombardment of atoms onto the target and collision kinematics is also carefully controlled externally by the specific experimental design. The damage caused by point defects (vacancies, interstitials etc.), any extended or complex defects (clusters, dislocation etc.) or unwanted contamination or impurities added during the various device fabrication processes play a vital role in making the TFT/LCD building blocks functional/dysfunctional and efficient/inefficient. The unnecessary and uncontrolled damage near the surface or device region badly impacts the electrical characteristics of the device (such as I-V operational trend, drain and transfer curves, sensors response time etc.) and physical electronic parameters (such as conductivity and electron/hole motilities) be creating meta-stable defects traps leading to a large leakage current and lower life time of the carriers responsible for the efficient functioning of the device. PIXE measurements confirm that our designed implantation experiments do not create any such issues while translating the process into a junction based full cycle TFT/LCD like unit device. There are no traces of 'Si' or 'He' or 'Si-He complex' piled up as residual dopants/ cluster defect (agglomerated on anti-sites or start acting as impurities/defect centers) within the active device region of the substrate. The minor traces of 'Ca', 'Zr' and 'Ti' are process based contamination with a very low content (and that too with progressively lower yield with higher annealing temperatures as observed in Fig. 6) and hence unable to become profound in affecting the electrical characteristics of transistor (the TFT/LCD building block) by creating any substantial defect trap levels. This suggests that the designed implantation process routine may successively be used with full confidence as a "substrate engineering" technique to produce a stable amorphized region for the TFT or LCD-like device application. The combined RBS and PIXE results in direct conjunction of TCAD SILVACO results (refer to Table 1, and Figures 1 and 2 together with numerous iterations and ATLAS based simulated electrical characteristics of TFT devices which are not shown here) also suggest that the variation of the He^+ implant doses within acceptable fluence regions of 1×10^{13} ion/cm^2 and 1×10^{16} ion/cm^2 play a decisive role in the adjustment of the controllable damage extent (using annealing as process routine for subsequent process steps where a higher or lower value of such damage extent is required). The medium to slightly higher concentration of interstitials up to a certain depth (owing the fact that helium ions with varying doses are impinged onto the n-type Si substrate at a given energy) can be achieved and controlled with respect to the derived physical or electrical device characteristics

The effect of helium implantation with theses carefully chosen parameters is important to create a "depth specific", "controlled and almost uniform damage profile" acting as an amorphized layer which may be reversible to crystalline undergoing the annealing process, if required. The effect of Si-implantation is also vital to control the nature of aforementioned damage as it is evident that the persistent damage in this case is not of interstitials but vacancies, which play an important role while determining the desirable mobility of free carriers in the device at any given depth and process temperature. Thus both He^+ and Si^+ implants owing the roles of interstitials (with changing depth profile) and vacancies (with an ability to be mobile in great deal from surface to bulk and vice-versa) seem necessary to engineer the substrate in a way to alter the n type surface/near surface region into a pseudo-amorphized region highly suitable for TFT fabrication and consequent applications.

5. Conclusion

Ion implantation induced damage of silicon substrate particularly by implantation of Si^+ and He^+ ions into the Si-matrix is studied to evaluate and optimize the process control for TFT applications in Large Area Electronics and Display Sensing Technologies. The substrate engineering protocol was also utilized in the conventional TFT device design to ascertain its applicability in the manufacturing process. Post- process characterization tools such as RBS and PIXE were also focused to evaluate the scope of "precisely controlled" and "fully design optimized" substrate-engineered TFT devices in the real fabrication environment. The process-altered design recipe of such TFT devices may prove to be useful in special reference to the control on the efficiency and response specific physical characteristics of the devices used in display sensors and consumer electronics industry.

References

[1] Foster P J, Doylend J K, Mascher P, Knights A P and Coleman P G 2004 *Applications of Defect Engineering to the Fabrication of Silicon-based Photonic Devices* Proceedings of SPIE **5577** 683
[2] Gawlik G and Jagielski J 2008 *Nuclear Instruments and Methods in Physics Research* **266** 1307
[3] Yuan Z, Li D, Wang M, Gong D, Fan R, Yang D 2008 *Vacuum* **82** 1337
[4] Wagner S, Gleskova H, Cheng I, Wu M, 2003 *Thin Solid Films* **430** 15
[5] Kim S K, Cho S I, Choi Y J , Cho K S, Pietruszko S M , Jang J 1999 *Thin Solid Films* **337** 200
[6] Sakai M, Tsutsumi T, Yoshioka T, Masuda A, Matsumura H 2001 *Thin Solid Films* **395** 330
[7] Wu M and Wange S 2001 *Applied surface science* **175-176** 753
[8] Cuscun M, Bonfiglietti A, Carluccio R, Mariucci L, Mecarini F, Pecora A, Stanizzi M, Valletta A, Fortunato G 2002 *Solid-State Electronics* **46** 1351
[9] Mayer M 2003 *Rutherford Backscattering Spectrometry (RBS)* workshop on Nuclear Data for Science and Technology Trieste Italy 1
[10] Shibata S 2011 *Research on ion implantation damage of the materials for the next generation semiconductor device* (Japan: Graduate school of Maritime sciences, Kobe University)

IOP Conf. Series: Materials Science and Engineering **51** (2013) 012006 doi:10.1088/1757-899X/51/1/012006

Transient plane source (tps) sensors for simultaneous measurements of thermal conductivity and thermal diffusivity of insulators, fluids and conductors

Asghari Maqsood[1], M Anis-ur-Rehman[2]
[1]Centre for Emerging Sciences, Engineering and Technology (CESET), Islamabad, Pakistan
Email: tpl.qau@usa.net, maqsoodasghari@gmail.com

[2]Applied Thermal Physics Laboratory, Department of Physics, COMSATS Institute of Information Technology, Islamabad, Pakistan.
Email: marehman@comsats.edu.pk

Abstract. Thermal conductivity and thermal diffusivity are two important physical properties for designing any food engineering processes[1]. The knowledge of thermal properties of the elements,compounds and different materials in many industrial applications is a requirement for their final functionality. Transient plane source (tps) sensors are reported[2] to be useful for the simultaneous measurement of thermal conductivity, thermal diffusivity and volumetric heat capacity of insulators, conductor liquids[3] and high-T_C superconductors[4].The tps-sensor consists of a resistive element in the shape of double spiral made of 10 micrometer thick Ni-foils covered on both sides with 25 micrometer thick Kapton. This sensor acts both as a heat source and a resistance thermometer for recording the time dependent temperature increase. From the knowledge of the temperature co-efficient of the metal spiral, the temperature increase of the sensor can be determined precisely by placing the sensor in between two surfaces of the same material under test. This temperature increase is then related to the thermal conductivity, thermal diffusivity and volumetric heat capacity by simple relations[2,5]. The tps-sensor has been used to measure thermal conductivities from 0.001 $Wm^{-1}K^{-1}$to 600 $Wm^{-1}K^{-1}$ and temperature ranges covered from 77K- 1000K. This talk gives the design, advantages and limitations of the tpl-sensor along with its applications to the measurementof thermal properties in a variety of materials.

1. Introduction

Recently a new transient technique has been proposed for thermal transport studies of solid materials' and a general theory of this transient plane source (TPS) technique is outlined below. The general theory is followed by proposed approximations for arrangements, which can be referred to as "hot square" and "hot disk". It is well known that the thermal conductivity as well as the thermal diffusivity of solids varies extensively depending on the structure, density, porosity, electrical conductivity, etc., of different materials. Frequently these properties also exhibit a strong dependence on temperature and pressure.

The motive behind the development of the TPS technique has been to cover as large ranges of the transport properties as possible and at the same time be able to apply the technique to a large number of different materials. It is also clear that maximum applicability as well as convenience can be achieved by the use of a "resistive element" both as heat source and temperature sensor. A transient plane source element used both as a heat source and a temperature sensor may consist of a pattern of a thin layer of an electrically conducting material, the temperature coefficient (TCR) of which is such that the temperature increase of the element can be precisely deduced from recording of its resistance. In case a metal foil is used as a TPS element it may be supported on both sides by thin insulating layers. Such an arrangement also makes it possible to apply the TPS element for thermal conductance studies of electrically conducting materials. It should also be noted that we will be discussing TPS elements with approximately the same overall length-to-width dimensions. The reason for this particular choice of pattern is the desire to keep the sample dimensions as small as possible.

2. THEORETICAL BACKGROUND

The TPS method considers the three-dimensional heat flow inside the specimen. The specimen is regarded as an infinite medium. The technique uses a ''resistive element'' (the TPS element) both as a heat source and temperature sensor. The time-dependent temperature increase $\Delta T(t)$ gives rise to a change in the electrical resistance, $R(t)$, of the conducting pattern shown in Fig. 1, and may be expressed as

$$R(t) = R_0(1 + \alpha\,\overline{\Delta T(t)}) \tag{1}$$

Here R_0 is the resistance of the TPS element before transient recording has been initiated, α is the temperature coefficient of resistivity (TCR), and $\overline{\Delta T(t)}$ is the mean value of the time-dependent temperature increase of the TPS element.

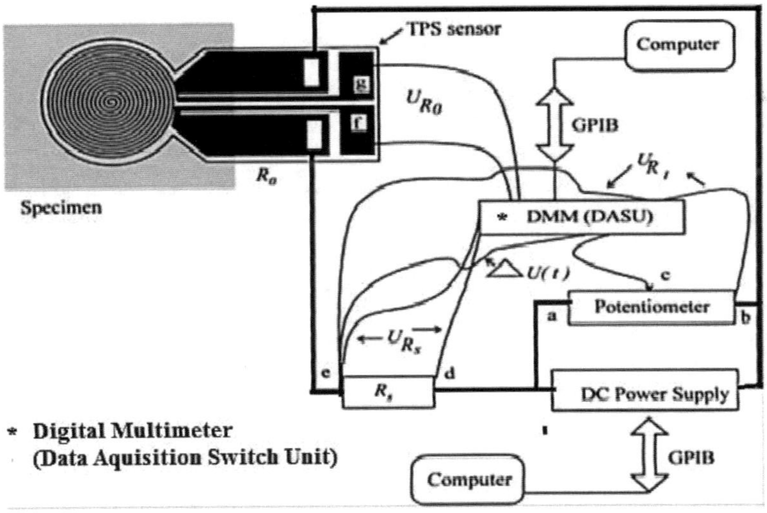

Figure 1.Principal electrical circuit for the modified bridge arrangement, showing TPS sensor (R_0) with specimen, standard resistance (R_s), potentiometer, dc power supply, DASU, and points of measuring voltages.

For convenience and as an experimental requirement, to obtain the thermal conductivity and thermal diffusivity simultaneously, the mean temperature change of the sensor is defined in terms of the dimensionless variable τ, where

$$\tau = \left(\frac{t}{\theta}\right)^{\frac{1}{2}} \tag{2}$$

and

$$\theta = \frac{r^2}{a} \tag{3}$$

Where$t$is the time measured from the start of transient heating, a is the thermal diffusivity of the specimen, θ is the characteristic time, and r is a constant that represents a measure of the overall size of the resistive pattern (=radius of hot disk).

3. EXPERIMENTAL ARRANGEMENT

The bridge circuit used is shown in figure. The electrical circuit consists of an HP 6633B dc power supply, R_s-standard resistance, potentiometer, and digital data acquisition switch unit (DASU) Agilent 34970A with internal DMM and TPS element R_0. DASU is used to measure quasi simultaneous voltages at the points d and e (U_{Rs}), f and g (U_{R0}), and b and e (U_{Rt}) of the circuit before and after the transient run. Also, the voltage variations $(\Delta U(t))$ (200 measurement points) across the bridge circuit during the transient run are measured by DASU. The dc output power of the power supply unit that is used for balancing the bridge and heating the TPS element is controlled through the computer. For balancing the bridge, a very low current is supplied to the circuit so that it works for balancing but does not actually heat the TPS sensor. The modified bridge circuit has many plus points as compared to the earlier used bridge, like minimal circuit components, reduced time between two consecutive measurements, better accuracy, high speed, better reproducibility and least error in results.

4. RESULTS AND DISCUSSION

With the purpose of demonstrating that the TPS technique using the modified setup could be used down to the temperature of liquid nitrogen, measurements on fused silica, carbon steel, and silver chloride crystals were performed. The main aim of the whole study is to develop a simple system providing higher accuracy that can be used for the measurement of the thermal transport properties of high-T_c superconductors. After verifying the performance of the setup, the superconducting samples with nominal composition $Bi_{1.6}Pb_{0.4}Sr_{1.6}Ba_{0.4}Ca_2Cu_3O_Y$ were studied. The dimensions of the four types of specimens and radii of the TPS sensor used are given in Table I.

Table 1.Dimensions of Test Specimens, along with Sensor Radii, Probing Depth, and τ_{max}

S. No	Test specimen	Diameter (±0.05)mm	Length (±0.05)mm	Sensor radius (±0.001)mm	Probing depth (±0.001)mm	τ_{max}
1	Fused Silica	(a) 35.05	(a) 20.30	(i) 9.734 (sensor 1)	(i) 8.645	(i) 0.50
		(b) 35.00	(b) 20.35	(ii) 3.300 (sensor 2)	(ii) 6.500	(ii) 0.98
2	Carbon Steel (SS215/3)	(a) 38.15	(a) 19.70	9.734	15.260	0.78
		(b) 38.25	(b) 19.50			
3	Silver Chloride crystals	(a) 25.40	(a) 4.30	3.300	3.972	0.60
		(b) 25.40	(b) 4.25			
4	$Bi_{1.6}Pb_{0.4}Sr_{1.6}Ba_{0.4}Ca_2Cu_3O_y$ Superconductor	(a) 28.01	(a) 11.12	3.300	7.473	0.41
		(b) 28.02	(b) 11.31			

The measurements are carried out as a function of temperature for all four specimens. Due to the availability of data for the thermal conductivities of the studied specimens from other sources (with different techniques), comparisons of thermal conductivities are made.

4.1. Fused Silica

Data for the silica specimens were taken with sensors having radii of 9.734±0.001 mm (sensor 1) and 3.300±0.001 mm (sensor 2). The variation of the thermal conductivity with temperature is shown in Fig. 2 and the results are tabulated in Table II. The results obtained are compared with the recommended experimental values and also with the previously calibrated bridge circuit. The differences do not exceed 1% at any temperature. Previously reported differences were 1% at room temperature and 4% at low temperatures. These results show that the performance of ATPS (Advantageous Transient Plane Source - Specific method used in the present study) has been improved.

The measurement times used were 10 to 20s for sensor 1 and 10s for sensor 2. The currents used for balancing were in the range of 1.5 to 3 mA and for measurements, the current values ranged from 149 to 325 mA. Balancing currents and measurement currents decrease as the temperature is increased as the resistance of the TPS sensor and the connecting leads changes with temperature.

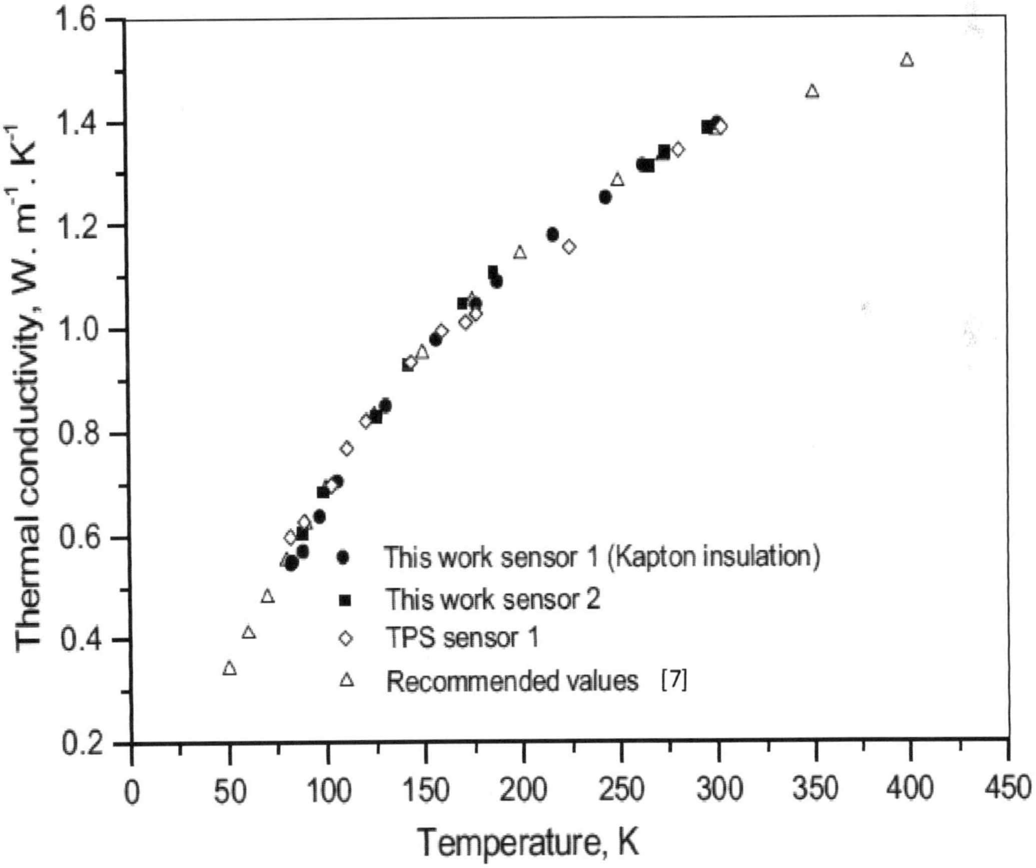

Figure 2. Variation of the thermal conductivity of fused silica samples withtemperature.

IOP Conf. Series: Materials Science and Engineering **51** (2013) 012006 doi:10.1088/1757-899X/51/1/012006

4.2. Carbon Steel

Carbon steel specimens, SS215/3, obtained from Analytical Standards AB Sweden, were analyzed from room temperature down to the boiling point of liquid nitrogen. A plot of the thermal conductivity variation with temperature and composition of the specimen is shown in Fig. 3, and the results are tabulated in Table III. A similar trend has been observed in an earlier study for an almost similar composition. They made measurements from room temperature to 573 K. The values for thermal conductivity are in agreement with this report for the same temperature range. Detailed comparisons are made with the data from an earlier measurement. The thermal conductivity values decrease at low temperatures due to the reduction of the electronic contribution to conduction.

4.3. Silver Chloride Crystal

Silver chloride crystals are a useful material for deep IR applications where sensitivity to humidity is a problem. A major use is in the manufacture of small disposable cell windows for spectroscopy, known as a "minicell". The structure of these crystals is cubic, fcc.

For the AgCl crystals, only sensor 2 was suitable. Fig. 4 shows the variation of thermal conductivity with temperature for the silver chloride crystals, and the results are tabulated in Table IV. A very good match is observed for the available recommended data, and also the thermal conductivity values are extended down to liquid nitrogen temperature. These data will be helpful for application to optical windows used at low temperatures. The thermal conductivity values of AgCl are proportional to T^{-1} in the temperature region of the measurement. The detailed calculations will be published elsewhere.

Table 2.Thermal Conductivities (in $W \cdot m^{-1} \cdot K^{-1}$) of Fused Silica from Other Sources andfrom Both Sensors Used in the Present Work

S. No.	*T* (K)	(i)[a]	(ii)[b]	Sensor 1	Sensor 2
1	50	0.34±0.03	–	–	–
2	60	0.41±0.03	–	–	–
3	70	0.48±0.03	–	–	–
4	80	0.55±0.03	–	–	–
5	82	–	0.595±0.020	0.5542±0.0007	–
6	88	–	–	0.6041±0.0008	0.6061±0.0008
7	89	–	0.625±0.021	–	–
8	90	0.62±0.04	–	–	–
9	97	–	–	0.6744±0.0009	–
10	100	0.69±0.03	–	–	0.6840±0.0009
11	103	–	0.696±0.023	–	–
12	106	–	–	0.7048±0.0010	–
13	111	–	0.768±0.024	–	–
14	121	–	0.821±0.024	–	–
15	125	0.83±0.03	–	–	0.8287±0.0011
16	131	–	–	0.8501±0.0011	–
17	144	–	0.933±0.027	–	0.9306±0.0013
18	150	0.95±0.03	–	–	–
19	157	–	–	0.9782±0.0013	–
20	160	–	0.995±0.028	–	–
21	172	–	1.010±0.028	–	1.0471±0.0014
22	175	1.05±0.04	–	–	–
23	177	–	1.026±0.025	1.0465±0.0014	–
24	188	–	–	1.0899±0.0015	1.1077±0.0015
25	200	1.14±0.04	–	–	–
26	204	–	–	–	1.1186±0.0015
27	217	–	–	1.1791±0.0016	–
28	225	–	1.155±0.023	–	–
29	244	–	–	1.2517±0.0017	–
30	250	1.28±0.04	–	–	–
31	263	–	–	1.3144±0.0018	–
32	266	–	–	–	1.3127±0.0018
33	273	1.33±0.04	–	–	1.3391±0.0018
34	281	–	1.342±0.026	–	–
35	296	–	–	–	1.3873±0.0019
36	300	1.38±0.04	–	–	–
37	301	–	–	1.3940±0.0019	–
38	303	–	1.386±0.027	–	–
39	350	1.45±0.04	–	–	–
40	400	1.51±0.04	–	–	–

[a] Touloukian and Powell, 1970 [7]

[b] Calibrated bridge by Maqsood et al., 1996 [6]

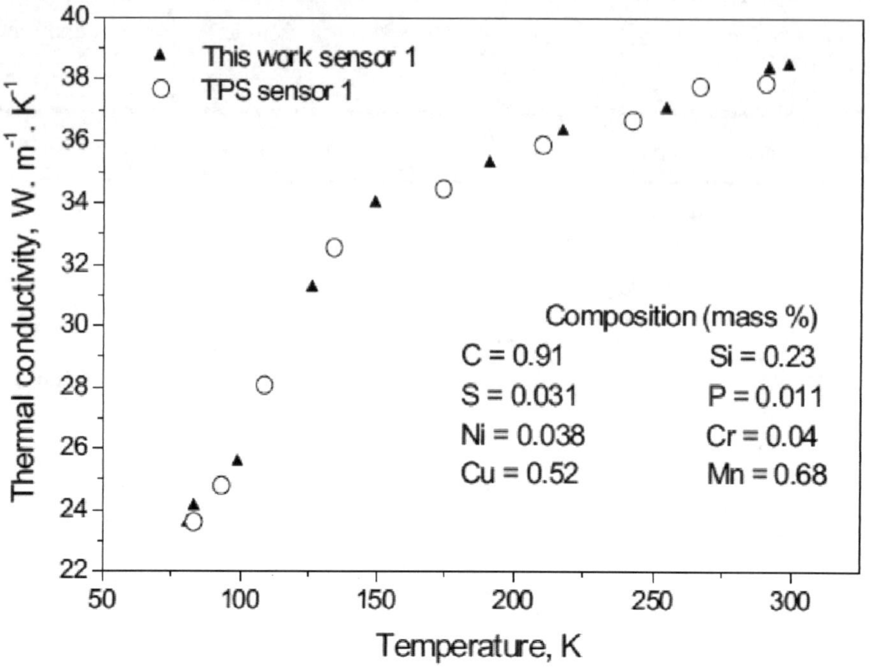

Figure 3. Variation of the thermal conductivity with temperature for the stainless steel No. 215/3 0.9% carbon

Table 3.Thermal Conductivity (in W·m^{-1} ·K^{-1}) of Stainless Steel from Circuit Used Previously and the Present Work

S. No.	T (K)	TPS [6]	This work
1	81	–	23.595±0.033
2	83	23.62±0.81	24.120±0.034
3	93	24.81±0.74	–
4	99	–	25.553±0.036
5	109	28.08±0.78	–
6	126	–	31.248±0.044
7	134	32.55±0.80	–
8	149	–	33.973±0.047
9	174	34.45±0.81	–
10	191	–	35.292±0.049
11	210	35.88±0.72	–
12	217	–	36.327±0.051
13	242	36.70±0.73	–
14	254	–	37.068±0.052
15	266	37.79±0.75	–
16	290	37.91±0.75	–
17	298	–	38.495±0.054

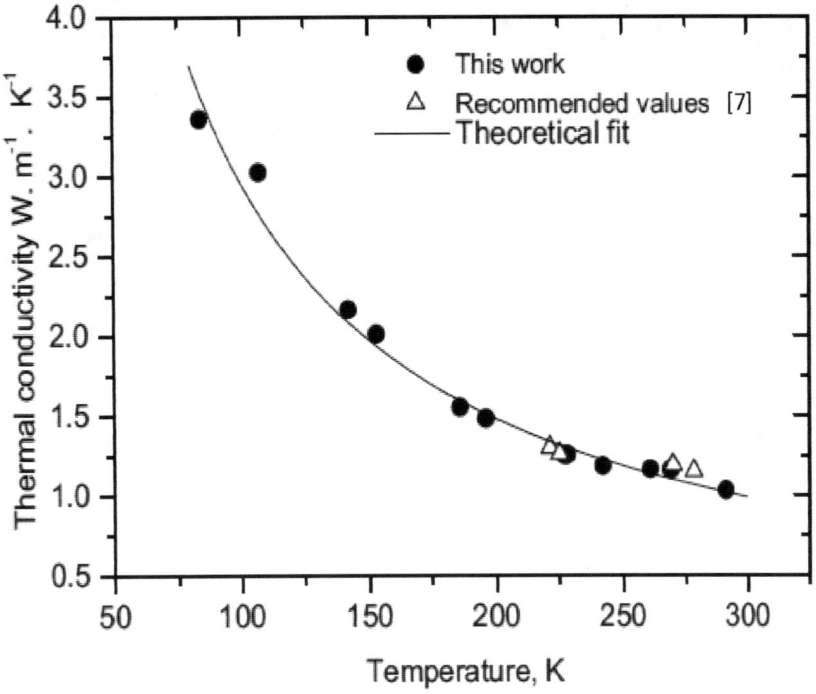

Figure4.Variation of the thermal conductivity with temperature for AgCl crystals. Solid line is $\lambda = cT^{-1}$ fit, where c is a constant.

IOP Conf. Series: Materials Science and Engineering **51** (2013) 012006 doi:10.1088/1757-899X/51/1/012006

Table 4.Thermal Conductivity (in $W \cdot m^{-1} \cdot K^{-1}$) of Silver Chloride Crystals from Other Sources and the Present Work

S. No.	T ($\pm$1 K)	Other sources [7]	This work
1	84	–	3.3612$\pm$0.0047
2	107	–	3.0305$\pm$0.0042
3	142	–	2.1661$\pm$0.0030
4	153	–	2.0121$\pm$0.0028
5	186	–	1.5514$\pm$0.0022
6	196	–	1.4823$\pm$0.0021
7	221	1.30	–
8	225	1.26	–
9	227	–	1.2511$\pm$0.0018
10	228	–	1.2560$\pm$0.0018
11	242	–	1.1842$\pm$0.0017
12	261	–	1.1643$\pm$0.0016
13	269	1.19	1.1612$\pm$0.0017
14	278	1.15[a]	–
15	291	–	1.0305$\pm$0.0014
16	325	1.09	–

[a] Crystran, United Kingdom.

4.4. Superconducting Samples

A Ba-doped, Bi-based, high-T_c superconductor was prepared by a solidstate reaction method. The nominal composition used was $Bi_{1.6}Pb_{0.4}Sr_{1.6}Ba_{0.4}Ca_2Cu_3O_y$. Large-sized samples (diameter~28 mm and length ~11 mm) are investigated for thermal transport properties. The size of thesamples used for the thermal conductivity measurements was indeed thelargest reported for this kind of measurements. Our thermal conductivityresults are plotted in Fig. 5 and are tabulated in Table V. The sampleshows the same general trend and at $T > T_c$ the thermal conductivity isslightly decreasing, but at $T < T_c$, it is enhanced.

Because the compound is metallic above T_c, heat can be conducted by both phonons (λ_{ph}) and electrons (λ_e). Thus, the largest contribution to the total thermal conductivity (λ) above T_c is due to lattice conduction. The sharp up-turn of the measured thermal conductivity and the further increase for $T < T_c$ has three different hypotheses: (i) only the phonons are contributing to this increase (phonon approach); (ii) only the electron contribution is responsible (electron approach); and (iii) both phonon and electron contributions are causing the increase (electron+ phonon approach).

It should be noted that the uncertainties in thermal conductivity values of our data are within 1 %. Comparing the results between different laboratories, one notes that the thermal conductivity depends on a particular sample preparation process, while the temperature dependence of the conductivity is similar for all the samples[6-8]. So the order of magnitude of the thermal conductivity is comparable to the results obtained by different authors[9], measured by steady-state methods. Further detail of this material has been discussed by Rehman and Maqsood[10].

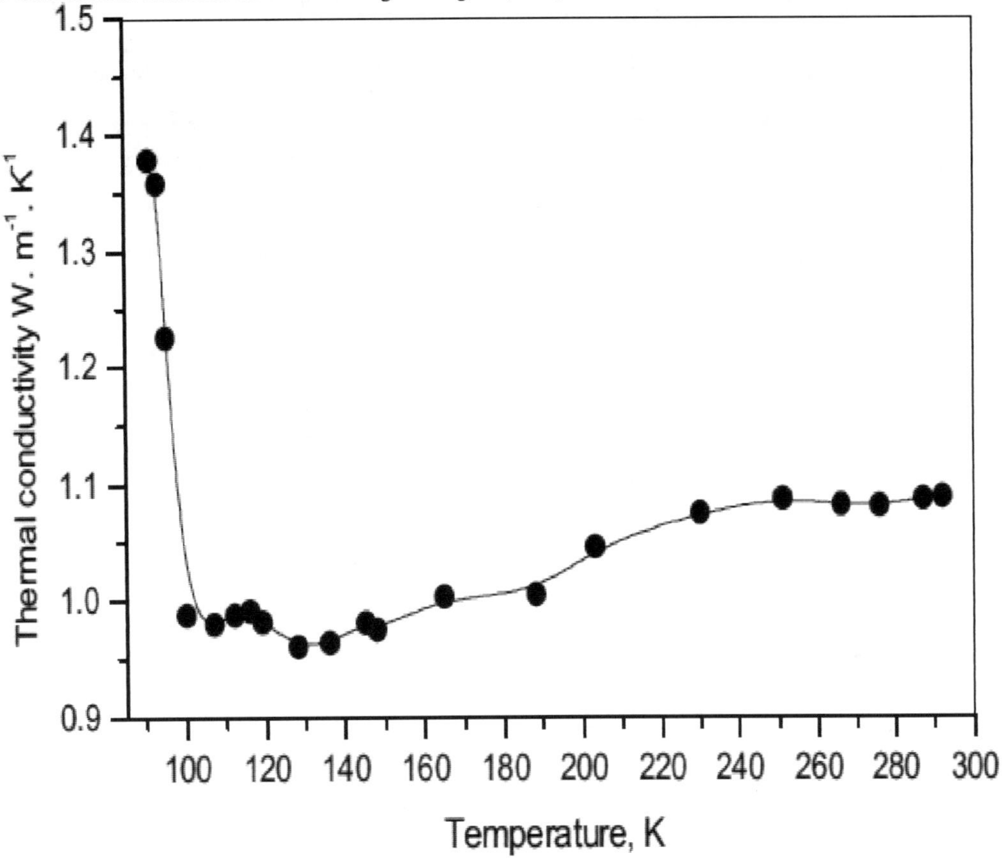

Figure 5.Variation of the thermal conductivity with temperature for the superconducting samples

Table V. Thermal Conductivity (in W·m−1 ·K−1) of the Superconducting Samples

S. No.	T (±1 K)	Thermal conductivity
1	91	1.3791 ±0.0013
2	93	1.3582±0.0013
3	95	1.2263±0.0017
4	100	0.9883±0.0014
5	107	0.9802±0.0014
6	112	0.9879±0.0014
7	116	0.9917±0.0014
8	119	0.9821 ±0.0013
9	128	0.9602±0.0013
10	136	0.9638±0.0013
11	145	0.9808±0.0014
12	148	0.9753±0.0014
13	165	1.0032±0.0014
14	188	1.0051 ±0.0014
15	203	1.0463±0.0015
16	230	1.0752±0.0015
17	251	1.0871 ±0.0015
18	266	1.0820±0.0015
20	276	1.0811 ±0.0015

5. SUMMARY AND CONCLUSIONS

An improved bridge for the transient plane source has been validated with measurements on fused silica, carbon steel, silver chloride and superconducting specimens. At any temperature, the deviation in the evaluated results of the thermal conductivity is less than 2% when compared with recommended values. With the present arrangement, the bridge circuit components are reduced, the temperature drift of the specimens is very well compensated, and the time of relaxation (time between two reliable readings) is minimized.

6. REFERENCES

[1] Huang L, Liu,2009 LS: J. Food Engineering 15179
[2] Gustafsson,1991SE: Rev. Sci. Instrum. 62797
[3] Maqsood A, et.al:1994 International J. Energy Research 18777
[4] Rehman M A, Maqsood A:2003 International Journal of Thermo Physics 24867
[5] Maqsood A:2012 Key Engineering Materials 510-511
[6] Maqsood M, Arshad M, Zafarullah M, Maqsood A: 1993J. Superconduct. Sci.Technol. 9321
[7] Touloukian YS, Powell RW: Thermal Conductivity, Nonmetallic Solids, Vol. 2, Thermophysical Properties of Matter (IFI/Plenum, New York, 1970)
[8] Ginsberg DM: High Temperature Superconductivity (World Scientific Publishing, London)
[9] Uher C:1990 J. Supercond. 3337
[10] Rehman MA, Maqsood A:2003 International Journal of Thermophysics24867

Advanced Data Communications for Downhole Data Logging and Control Applications in the Oil Industry

C T Spracklen[1]
Intérim Dean of Science & Engineering, Habib University, Karachi, Pakistan

Tariq Aslam
VP Ericsson CSI, UAE

Abstract. We present details of 'Mercury', a high-speed downhole communications system that utilizes the (metallic) wall of a gas or oil pipeline or a drill 'string' as the communications 'channel' to control or monitor equipment or sensors used in the oil industry. Conventional downhole communication systems typically use 'mud pulse' telemetry for 'Measurement While Drilling' (MWD) operations. Current mud pulse telemetry technology offers bandwidths of up to 40 bit/s. However the data rate drops with increasing length of the wellbore and is typically as low as 1.5 bit/s - 3.0 bit/s at a depth of 35,000 ft. - 40,000 ft. The system described, by contrast, offers data rates of several megabits per second over distances of many kilometres and uses Orthogonal Frequency Division Multiplexing (OFDM) coupled with Wideband Frequency Division Multiple Access (W-CDMA). This paper presents details of our system; results of several trials undertaken on actual gas pipelines in the United Kingdom will be presented at the Conference.

Keywords: Downhole communications, MWD, OFDM, DSSS, W-CDMA, FPGA.

1. Introduction

Harsh communications environments, such as those encountered in downhole applications, present special challenges for improving data transmission technology and have precluded the use of existing techniques. Some solutions have explored pulsed communication through fluids and inductive coupled communication between drilling pipes. Pulsed communication has the drawback of a long delay between the sending and receiving of data and inductive coupled pipes suffer from loss of data integrity and signal strength in the past.

Mercury is a high-speed downhole communications system that utilizes the (metallic) wall of the pipe as the communications channel. It has been recently demonstrated in a series of tests and trials – at the Theddlethorpe Gas Refinery, near the Lincolnshire Coast in England, at the Montrose RGIT training well in Scotland and in an extended six-week trial between the St Fergus and Cruden Bay gas terminals, also in Scotland, during which time communication speeds of several megabits per second were consistently achieved.

Mercury uses a combination of two modulation techniques to handle the downhole environment. To combat multipath, where signals are reflected from discontinuities in the electrical impedance of the pipe and so multiple 'echoes' of the signal arrive at the receiver, Mercury uses Orthogonal Frequency Division Multiplexing (OFDM) [1]. In OFDM the communications bandwidth is subdivided into many parallel sub-channels. In Mercury there can be as many as 1024 parallel frequency channels. In addition Mercury uses Wideband Frequency Division Multiple Access (W-CDMA) [2], a form of Direct Sequence Spread Spectrum (DSSS) [3] modulation that allows it to dynamically adapt to changing channel conditions and to trade off data rate for signal-to-noise ratio (SNR), as well as to accommodate multiple downhole sensors or equipment.

---

[1]The authors are grateful to PACT (Engineering) Scotland and the ITF (Industry Technology Facilitator), Scotland for supporting this project.

Figure 1 – Typical sub-sea environment for Measurement-While-Drilling

Orthogonal Frequency Division Multiplexing (OFDM) is a method of transmitting data over a range of frequency bands. The main principle of OFDM is that it has a set of sub-carriers within one particular frequency range. This main feature is exploited by making the adjacent sub-carriers orthogonal to each other. This produces two main benefits: The first is that the symbol period is increased, as each sub carrier holds only one symbol; this means that the length of the whole frame is in fact the length of the symbol, thus reducing inter-symbol interference (ISI). The other main advantage is that the guard interval, that is typically used to protect frequencies from interfering with each other, can be removed as a result of the orthogonal properties of the carriers. This means that OFDM can produce data rates close to the Nyquist rate and is thus seen to be a highly efficient means of data transmission.

Processing is carried out using a specially designed miniature low-power processor card that contains an Atmel 32-bit microcontroller, RAM, Flash memory and an FPGA (that undertakes all the intensive DSP operations required for the OFDM and W-CDMA signal processing) [5]. A bespoke version of an open source real-time operating system (RTOS) is used and programming is carried out using Keil C for the microcontroller and a version of VHDL for the FPGA.

Without the use of the FPGA, it would not be possible for Mercury to carry out the compute-intensive signal processing operations needed to process the OFDM and CDMA protocols in real time. Fast Fourier Transforms (FFTs) of data sets comprising several thousand elements are required at the channel symbol rate. An FPGA is ideally suited to performing operations that are inherently parallel in nature – such as the FFT – while the serial nature of computations carried out by a conventional microcontrollers mitigate against their use in such environments. The entire FPGA processing was simulated using the well-known mathematical software, MATLAB® using the Xilinx® FPGA System Generator and HDL (Hardware Description Language) Coder.

2. THE PIPELINE COMMUNICATIONS CHANNEL

2.1. Ideal and Real Communications Channels

An 'ideal' communications channel would pass all electrical signals of any frequency or amplitude without attenuating them and without adding noise or interference. Clearly such channels do not exist. In practice all communications channels attenuate signals and typically attenuate higher frequency signals to a greater extent than lower frequency signals. In addition, all communications channels add noise and, unless the channel is a theoretical model known as a transmission line, will reflect some of the signal, propagating from the transmitter to the receiver, back towards the transmitter. This effect results from 'discontinuities' in the electrical characteristics of the communications channel between the transmitter and receiver. It is these discontinuities that cause the propagating signal to be partially reflected back towards the transmitter.

In addition – and sometimes, but not always, as a consequence of the previous effect – most communication channels contain multiple physical paths between the transmitter and the receiver and, as a result, multiple copies ('echoes') of the transmitted signal will arrive at the receiver. This is known as 'multipath propagation'.

Multipath propagation is a common occurrence in digital communications and there are several ways in which it can be tackled. Relatively recently a modulation scheme known as OFDM – or orthogonal frequency division multiplexing – has been developed that is particularly resistant to the effects of multipath. OFDM is used in the Digital Audio Broadcasting (DAB) standard for digital radio [6].

We might expect the characteristics of a pipeline to be subject to considerable electrical discontinuities – causing multiple reflections and re-reflections – as well as multipath. Such electrical discontinuities would be likely to originate from the various joints, welds, valves, and junctions etc. that make up a typical pipe. As a result, OFDM might be expected to be the 'protocol of choice' for such a system, since it would seem to offer the best chance of combating these effects.

A further problem such a system faces – over and above those that have been discussed already – is that, in the case of MWD [4], the channel characteristics are not constant and change with time. This is a result of sensors and actuators moving, thus causing the length of pipe or drill string over which the signals have to propagate to vary. One might logically expect that, the further away from the transmitter such a system moves, the poorer the 'quality' of the channel will become.

There are several solutions to this problem. We could design the communications modulation scheme and protocols (including the error handling) to deal with the 'worst-case scenario'. But this inevitably implies that, when conditions are benign, we are operating in a sub-optimum environment because we have reduced the effective data rate to be below the Channel Capacity to ensure that all data arrive correctly.

A more intelligent approach would be to try to find a way in which we are able to change the operation of the communications system to dynamically adapt to the varying communications channel conditions. We need to find a way to modify the communications modulation scheme, protocols and/or error handling to reduce the effective data rate to below the channel capacity.

In Mercury CDMA provides the dynamic response to changing channel conditions; OFDM provides the resistance to multipath. CDMA employs a modulation technique known as 'Spread Spectrum', which allows us to dynamically 'trade off' the rate of data transmission with the channel characteristics. The first author has worked in the field of spread spectrum for many years and has many publications and patents in this area.

2.2. Measuring the channel characteristics

In order to understand the electrical characteristics of a typical gas pipeline, tests were carried out at the Conoco Theddlethorpe gas refinery, which is situated in the south of the UK. We wanted to understand how the electrical characteristics of the pipeline vary with distance. This information can be used to select the parameters for implementing the CDMA part of the modulation scheme. At Theddlethorpe we had access to a 50km length of active gas pipeline, between the Theddlethorpe gas terminal and the Humber Refinery, that was buried, but where cathodic anti-corrosion access points were available at frequent intervals along its length. Figure 2, below, shows the transmit antenna being attached to the Humber Oil Refinery Export Line, which carries condensate from Theddlethorpe the 50 km to the Humber Oil Refinery.

In the trials at Theddlethorpe, the propagation characteristics were measured from antennas attached to the outside of the pipeline. For many other situations however, such as MWD, the equipment would be communicating to actuators or sensors contained within the well-bore itself. To facilitate the measurement of channel characteristics for such situations, a separate set of trials was carried out at the RGIT Training Well installation, near Montrose in Northern Scotland. This installation is used for training oil workers and forms a part of the necessary certification that anyone working in the oil industry – either on-shore or off-shore – needs to have. The well that we had access to was shallow by oil industry standards (2500'), but it provided a safe and flexible and, more importantly, non-production environment in which we were able to carry out extensive measurements.

Figure 2 – Attaching the antenna to the Humber Oil Refinery Export Line

A picture of the RGIT Montrose installation is shown below in Figure 3. The downhole module was enclosed in a stainless-steel pressure housing and was lowered down the well bore in increments of 50 ft. using an isolated wireline [7]. At each location the data rate, BER, SNR etc. were measured and logged using a commercial data logger at the surface connected to the topside transmit/receive module. At the well bottom, the downhole module was detached from the wireline, which was retrieved, in order to guard against the possibility that the wireline itself was acting as a conduit for the signals.

Figure 3 – The RGIT Training Well at Montrose in Northern Scotland

3. THE ST. FERGUS TO CRUDEN BAY TRIALS

3.1. Overview

3.1.1. Introduction

Figure 4 shows an aerial view of the St. Fergus Gas Terminal, as well as a schematic illustrating the relative positions of St. Fergus Gas Terminal, the Cruden Bay Gas Terminal and their interconnect to the North Sea oil fields. The equipment at both locations was also connected by high-speed mobile Internet to the project offices in Aberdeen, also in Northern Scotland. This allowed the experiment to be controlled and monitored remotely and for statistical data, illustrating the real-time performance of the pipeline communications network between the two gas terminals, to be observed. This was necessary because access to the gas terminals is strictly limited to authorized personnel and frequent visits, to reconfigure the equipment or collect data, were not practical.

3.1.2. The St. Fergus to Cruden Bay Trial

For the benefit of the project sponsors, these data were also made available in real-time on a dedicated web server, allowing them to 'log in' at will and note the key current network performance parameters – such as data rate, BER (bit error rate), SNR (signal-to-noise ratio) etc. The experiment ran continuously for a six-week period at which point it was terminated due to operational reasons.

Processing at both St. Fergus and Cruden Bay was performed using the bespoke processor board mentioned earlier, comprising the Atmel 32-bit processor and FPGA. A secondary processor card was employed as a dedicated interface to the 3G cellular data modems as well as to control the main Mercury processor card, consolidate and analyze the received data and to receive, interpret and implement commands sent over the cellular modem. Both installations – at St. Fergus and at Cruden Bay – were encased in ATEX approved enclosures. Since July 1st 2003 it has been mandatory under European law that all equipment for use in a potentially explosive atmosphere must conform to specific safety standards. An ATEX enclosure (from the French "ATmosphère EXplosible") provides the necessary protection for electrical equipment operating in such an environment. Since both St. Fergus and Cruden Bay are operational gas terminals, compliance with ATEX Directive 94/9/EC is mandatory.

During the trial, data rates of between 5 and 8 megabits per second were consistently observed and large files of data were repeatedly transferred back and forth to simulate the retrieval of data from pipeline sensors and the control of downhole actuators. Comprehensive statistical data were gathered for subsequent analysis and to guide the future 'tuning' of system parameters. These will be presented at the conference.

Although both the St. Fergus and Cruden Bay terminals are subject to significant EMI (electromagnetic interference), as a result of the extensive electrical machinery operating in both locations, no adverse effects were noticed during the six-week trial and there appeared to be no occurrence of diurnal or other periodic effects, as electrical machinery was switched on and off.

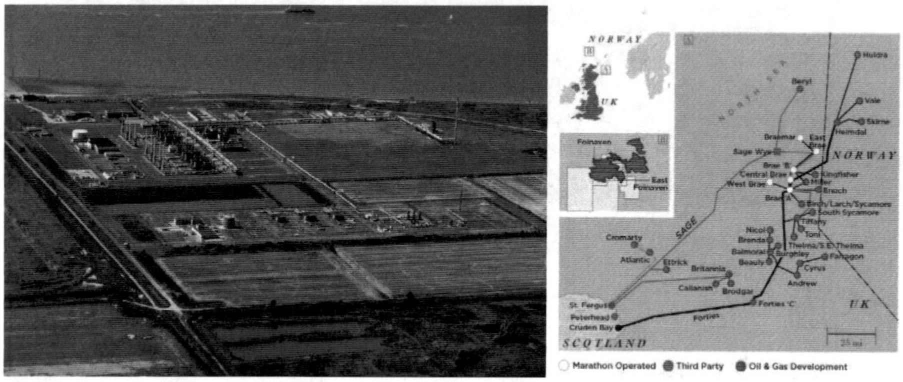

Figure 4 – An aerial view of the St. Fergus Gas Terminal in Northern Scotland and a schematic showing the interconnect

4. CONCLUSION

In this paper, we have described a downhole communications system that uses advanced communications protocols and the metallic drill string or pipe wall, to achieve data rates far in excess of those achieved using conventional techniques – such as mud-pulse telemetry. Comprehensive tests were undertaken to characterize the electrical characteristics of the pipe and drill strings at the Theddlethorpe Gas Refinery and the RGIT Training Well. A six-week trial between the St. Fergus and Cruden Bay Gas Terminals in Northern Scotland, consistently produced data rates of between 5 and 8 megabits per second. Further details of the communications protocols and results obtained will be presented during the conference. It is hoped to deploy the system in operational environments in the near future.

5. References

[1] Ramjee Prasad, OFDM for Wireless Communications Systems, Artech House Publishers (August 31, 2004)

[2] Andrew Richardson, WCDMA Design Handbook, Cambridge University Press (March, 2005)

[3] J. E. Hershey, Direct Sequence Spread Spectrum Techniques, Aegean Park Pr (June 1985)

[4] Norman J. Hyne, Nontechnical Guide to Petroleum Geology, Exploration, Drilling, and Production, 2nd Edition, PennWell Books, 2001

[5] Steve Kilts, Advanced FPGA Design: Architecture, Implementation, and Optimization, John Wiley & Sons, 18-Jun-2007

[6] Wolfgang Hoeg, Thomas Lauterbach, Digital Audio Broadcasting: Principles and Applications of DAB, DAB + and DMB, John Wiley & Sons, 03-Jun-2009.

[7] Sylvain Boyer, FranCois Coppens, Wireline Logging, Editions Technip (February 16, 1999)

IOP Conf. Series: Materials Science and Engineering **51** (2013) 012008 doi:10.1088/1757-899X/51/1/012008

Parallel Planar-Processed and Ion-Induced Electrically Isolated Future Generation AlGaN/GaN HEMT for Gas Sensing and Opto-Telecommunication Applications

S Ahmed[1,2], S H Bokhari[1], L A Khan[2], F Amin[1,2] and Z Hussain[3]
[1]Advanced Electronics Laboratory, Faculty of Engineering & Technology, International Islamic University, Islamabad - 44000, Pakistan.
[2] Centre for Emerging Sciences, Engineering & Technology, Islamabad-44000, Pakistan.
[3] Lawrence Berkeley National Laboratory, Berkeley, California 94720, USA.

E-mail: ahmed.shuja@iiu.edu.pk

Abstract. Ion-implanted AlGaN/GaN High Electron Mobility Transistors (HEMT) devices were studied thoroughly to look into the possibilities of enhancing efficiency for high-power and high-frequency electronic and gas sensing applications. A dedicated experimental design was created in order to study the influence of the physical parameters in response to high energy (by virtue of in-situ beam heating due to highly energetic implantation) ion implantation to the active device regions in nitride HEMT structures. Disorder or damage created in the HEMT structure was then studied carefully with electrical characterization techniques such as Hall, I-V and G-V measurements. The evolution of the electrical characteristics affecting the high-power, high-frequency and ultra-high efficiency gas sensing operations were also analyzed by subjecting the HEMT active device regions to progressive time-temperature annealing cycles. Our suggested model can also provide a functional process engineering window to control the extent of 2D Electron mobility in AlGaN/GaN HEMT devices undergoing a full cycle of thermal impact (i.e. from a desirable conductive region to a highly compensated one).

1. Introduction

The semiconductor device needs to be doped with impurities to generate electrons in a specific region. However, this procedure may cause the free carriers to slow down because they end up colliding with the impurities residing in the same layer that were used to generate them in the first place [1, 2]. High Electron Mobility Transistor (HEMT) is a smart device that was designed to resolve this contradiction. HEMT accomplishes this by using high mobility electrons generated using the heterojunction of a highly doped n-type AlGaN thin layer and non-doped GaN layer [1]. HEMT, by virtue of its efficiency, is widely used as an extremely low noise device in terrestrial and space telecommunications systems, radio telescopes, Direct Broadcasting Satellite television (DBS) receivers, car navigation receivers, photonic sensing and ultra-high efficiency gas and PH sensors [1, 3].

Gallium nitride (GaN) based HEMTs excel over competing technologies for high power and high frequency applications [1, 3]. GaN HEMTs have an order of magnitude higher power density and higher efficiency than those of silicon (Si) and gallium arsenide (GaAs) transistors, allowing a ten time size reduction for the same output power, while simultaneously saving material cost [3]. The wide band gap of GaN allows for rugged high voltage and high-temperature sensing applications extensively covering both commercial and military markets [1, 3].

AlGaN/GaN HEMTs with varying source-gate spacing are usually fabricated using a standard technology process developed in various labs [3]. One such laboratory tested and commercially viable scheme [3] is intentionally chosen for our experiments to evaluate the designed process in our study. This HEMT structure is shown in figure 1 [3].

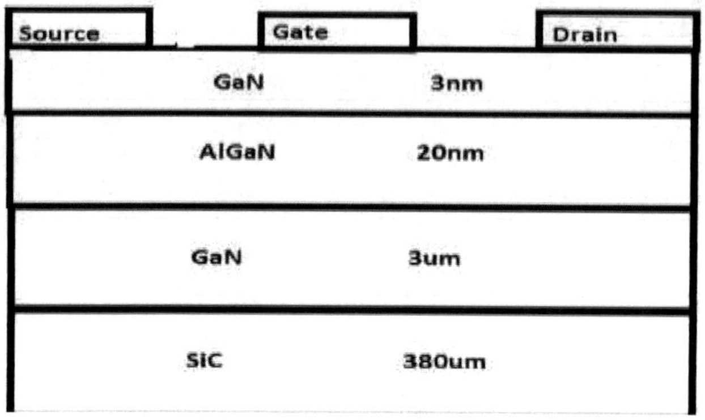

Figure 1. HEMT Structure (Commercially viable and Lab-scale qualified device: William J. Roman/NNIN IREU Site: Julich Forschungzentrum, Germany) [3]

The III-V semiconductors are currently employed in the fabrication of optoelectronic devices and integrated circuits (IC) for high frequency operations [4-16]. Electrical isolation of III–V semiconductors formed by introduction of a controlled concentration of point defects from light and heavy ion implantation is a well-established technique [6-16]. Besides the efficiency of this method to obtain highly isolated layers, it has the additional advantage of maintaining the planarity of the isolated structures [13-15]. Thermal processing of the crystal tends to remove ion beam induced damage and reduce the degree of compensation, while at the same time producing diffusion of the defect centers [8, 11-13]. Therefore thermal processing due to variable implantation temperature and post-implant annealing cycles is crucial to enable electrical and optical characteristics.

2. Experimental Conditions, Design and Details

The samples studied (*test device is intentionally chosen to be of the same design as used commercially [3, Figure 1] to independently evaluate the influence of high energy implantation on AlGaN/GaN layers and subsequent impact on the physical properties important to enable sensing, frequency and power related device operations*) were commercially grown by Metal Organic Chemical Vapor Deposition (MOCVD) on semi-insulating SiC substrate. The 2 Dimensional Electron Gas (2DEG) GaN channel was prepared with sheet carrier concentrations of $\boldsymbol{n}_s = 3.1\times10^{13}$ /cm^2 and $\boldsymbol{n}_s = 3.1\times10^{14}$ /cm^2. The ion implantation was carried out by Helium and Silicon ions irradiated at energy of 3000 KeV and 1000 KeV, respectively to place the peak of the damage distribution well inside the substrate as well as the GaN channel making an interface with the AlGaN barrier. A relatively flat damage distribution is created both in the top-layered GaN "cap" and AlGaN "barrier" so that a controlled amount of defects may be created at "interfaces" and "surface" for possible exploitation during the high power, high frequency and sensing applications. These implants were designed and modeled under the scope of this study and carried out at IBS Inc. USA with exact modeled parameters and dimensions. During implantation, the normal to the wafers was inclined at 15° with respect to the beam to minimize ion channeling. The selection of Helium and Silicon for damage studies for these HEMT test structures was crucial. Although the design studies included a range of ion species such as Zinc, Chlorine, Oxygen, Nitrogen etc. but the actual experiments were limited to one inert and lighter ion specie (Helium) and Silicon-a commonly used dopant in III-V devices. Silicon dopant is not known to create chemical impurity levels in AlGaN/GaN structures unlike Zinc, Chlorine, Oxygen and Nitrogen etc. The in-situ thermal effects caused by relatively high energy implantation due to beam heating and self-annealing of defects in HEMT devices were suspected to play an important role in affecting the device characteristics. Isolation implants were performed at different combined doses ranging from $5x10^{13}$ cm^{-2} to $2x10^{15}$ cm^{-2} at room temperature (RT). The doped regions under the contact areas were masked using the metal foil on top of those contacts. The resistivity and Hall measurements were carried out using Ecopia HMS3000 system, whereas HP 4156 parameter analyzer was used to evaluate the DC characteristics. Some of the measurements, implant schedules and annealing cycles were repeated and double checked at facilities in Berkeley National Laboratory, USA.

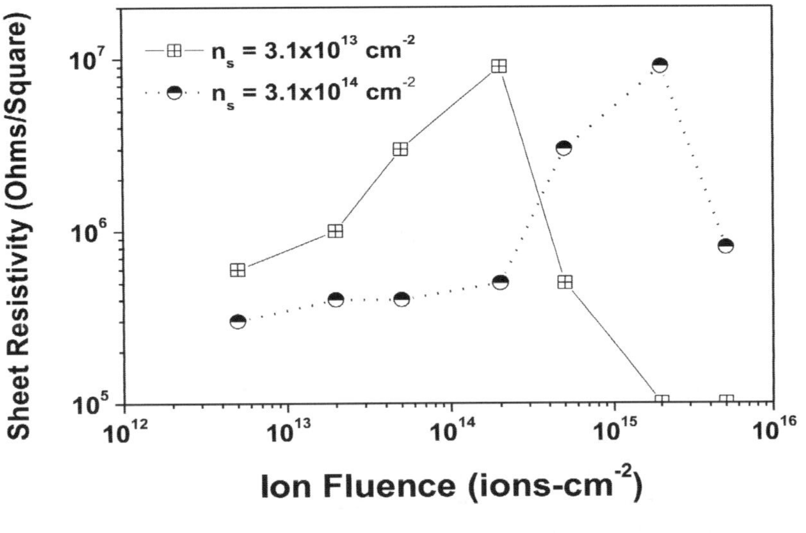

Figure 2. Evolution of sheet resistance with the increasing ion dose (experiments designed as similar as depicted in Reference 17 to qualify the model requirements)

3. Results, Analysis and Discussion

Utilizing the similar model as devised in reference 17, we conducted a very similar experiment to qualify the threshold dose model for III-N-V devices. Following reference 14, 16 and 17, a threshold dose for isolation, D_{th}, can be defined as the minimum ion dose at which sheet resistivity, *Rs*, reaches its maximum value. The presence of threshold doses can easily be observed in Fig. 2, where two distinct threshold doses exist for two different initial sheet carrier concentrations of GaN layers. It has already been demonstrated [14, 16, 17-20] that ion doses required for an effective isolation of GaAs devices have simple linear and reciprocal dependences on the values of initial free carrier concentration and ion-beam-produced atomic displacements, respectively. If n_{so} is the original sheet carrier concentration of the n-type GaN layers, D is the implanted dose to isolate the structure, n_{sD} is the measured sheet carrier concentration after the isolation implantation to dose D then R will be the carrier removal rate which is a function of mass and energy of the ion. The equation relating these parameters is as follows [17]:

$$nso = nsD + RxD \quad (1)$$

Since we do not intend to fully compensate the layers, optimum carrier removal rate is not considered while designing these experiments. The modelled parameters (using the Monte Carlo simulator called Stopping and Range of Ions in Matter [2] and SUSPRE) in terms of accumulated disorder and energy loss due to electronic and nuclear stopping are summarized in Table 1 for each case.

Table 1. The modelled physical parameters in response to the high energy ion implantation to the active device regions in nitride HEMT structure

PHYSICAL PARAMETERS	3000 KeV He+ into SiC at R_p (10.7 microns)	3000 KeV He+ into GaN Channel (3 microns)	1000 KeV Si+ into GaN Channel at R_p (1.45 microns)
Energy loss due to Electronic Deposition (electron volt/angstrom)	11.98	23	14.88
Energy loss due to Nuclear Deposition (electron volt/angstrom)	0.19	2.46	21
Accumulated Disorder %	7.3	1.9	15.1

The threshold doses in this experiments are $D = 2\text{x}10^{14}$ cm^{-2} and $D = 2\text{x}10^{15}$ cm^{-2} for the n-type GaN layers with initial channel sheet carrier concentrations of $3.1\text{x}10^{13}$/cm^2 and $3.1\text{x}10^{14}$/cm^2, respectively. Figures 3 and 4 illustrate the isolation characteristics (resistivity) for different starting material layers implanted with combined implantation at RT for a range of doses including the threshold doses in each case. Each data point is comprised of the cumulative isochronal annealing sequence for 50s period. For the D_{th} case in Figure 3, the stability of the isolation is restricted to temperatures below 400 °C. In the samples irradiated to doses higher than the threshold dose ($2.5D_{th}$) the isolation persists up to the temperature of 350 °C. The increase of the irradiation dose to $10D_{th}$ leads to the decrease of R_s because of hopping conduction in the highly damaged region. During post irradiation annealing there is a progressive increase of R_s with increasing temperature up to 300 °C, since the carrier hopping becomes less effective as the damage is repaired. The defects are removed by annealing above this temperature, which would correspond to the annealing of defect clusters or defect complexes, as already observed and reported in cases such as experiments done in reference 17 and corresponding references therein.

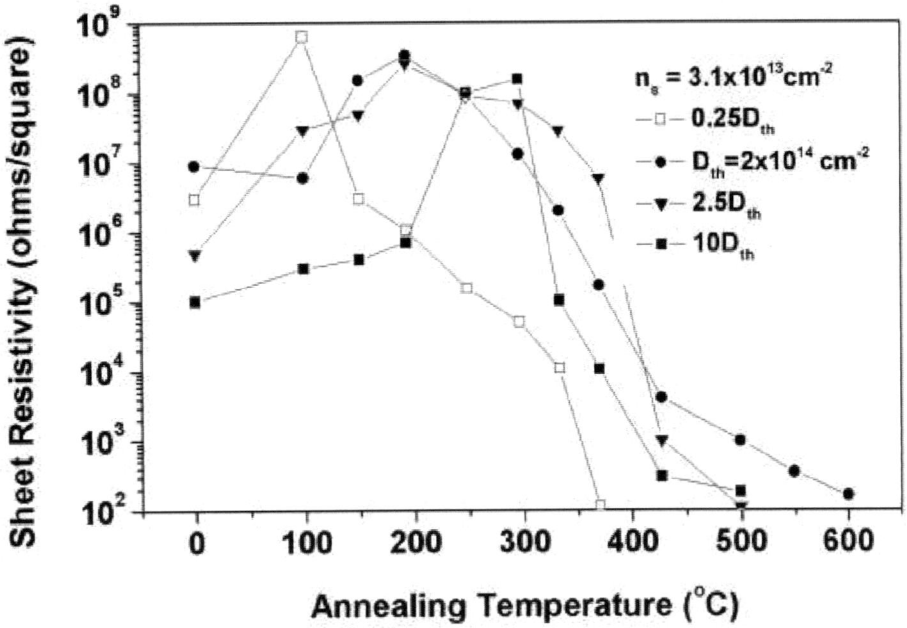

Figure 3. Carrier Removal characteristics (sheet resistance versus post-implant annealing temperature) for AlGaN/GaN device structures of initial sheet carrier concentration of 3.1×10^{13} cm^{-2} in GaN channel.

Figure 4 also explains that the devices of starting material with high sheet channel carrier concentration are reasonably isolated with threshold doses. Further higher doses, for example in the case of Figure 3, are not needed to obtain better isolation. The larger thermally stable window (till 500°C) achieved in the case of the results shown in figure 4, is sufficient for subsequent processing such as contact sintering. The evolution of sheet resistivity with varying annealing temperatures at certain ion fluence (variable doses with respect to the already calculated threshold doses in Figure 2) shown in Figures 3 and 4 follow the similar trend and behave almost identically to a similar finding in case of GaAs layers isolated with protons [20]. This is expected as the post-implant-annealed-isolation trends are reported to behave in a similar fashion for most of the III-V devices, particularly for damage-only ion species such as protons, helium and silicon [6-15, 20]. As part of our study, we aimed to employ the technique of ion beam engineering onto the scheme of commercially designed AlGaN/GaN HEMT structures. While using the modelled parameter of threshold dose during the ion implantation, it is experimentally shown above that one can make the layers highly resistive and hence the extent of mobile electrons can be reduced according to the requirements of the device operations. One can achieve a fully compensated layer either in capped layer, barrier layer, channel, buffer or substrate. The design and modelling of ion implantation parameters, such as in our experiments, provide us a flexibility to control the amount of mobile carriers within a certain layer or interface, which may in turn be utilized to control the high-power and high-frequency operations. The usage of high energy implantation (particularly the MeV range of incident energies) is known to produce special defect complexes due to the dynamic annealing during the process of implantation. Therefore, such thermal effects also contribute in the significant results achieved after the implantation in form of physical changes experienced by the lattice. With the scope of high electron mobility transistor's operation in mind, our proposed design of ion implantation parameters also provide an effective solution to improve the efficiency of device characteristics.

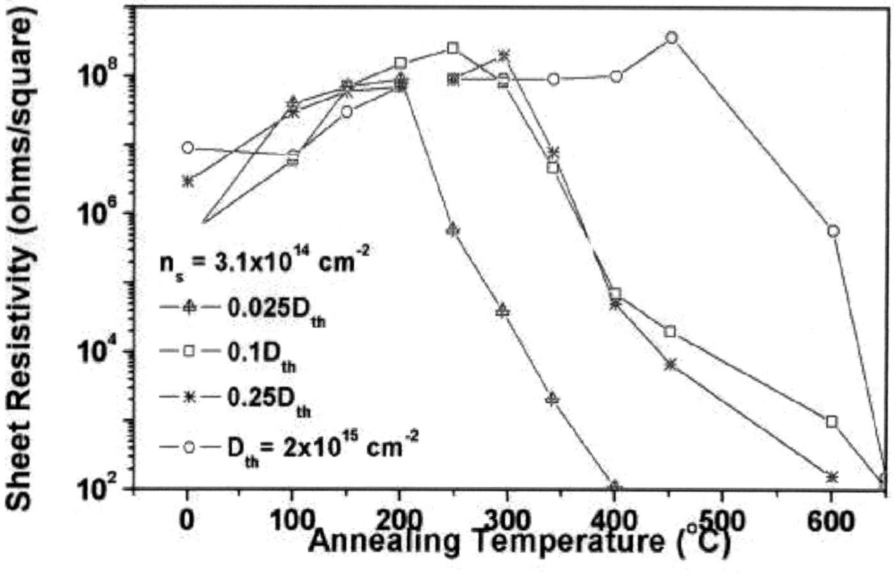

Figure 4. Presence of persistent electrical compensation: thermally stable process window for combined Helium and Silicon Implantation-compensated AlGaN/GaN device structures of initial sheet carrier concentration of $3.1x10^{14}$ cm^{-2} in GaN channel. These structures are implant-isolated at RT for a range of doses including the threshold doses in each case. Each data point is comprised of the cumulative isochronal annealing sequence for 50s period.

Being used in RF/Microwave amplification environment, nitride HEMT devices are considered to have efficiencies to operate both in low and high field carrier transport. This demands a high carrier density along with high carrier mobility even during the relatively high temperature environments. Persistent low sheet resistivity and high 2DEG mobility is desirable physical attribute of such devices. This can be only achieved if we fabricate the HEMT structure with (a) high spatially distributed conductivity of GaN channel, and (b) high quality of AlGaN/GaN interface. The higher strain and better carrier confinement on AlGaN/GaN interface may provide a desirable quality lower sheet resistivity as well as higher 2DEG mobility.

We suggest that by using the high energy ion implantation with a uniformly distributed damage distribution in the capped layer (near surface regions) and peak of the damage profile within the channel with a reasonable extent of damage therein may create such defects within the lattice which may confine the carriers at the interface effectively. The follow-up annealing at a certain temperature, while recovering the lattice after ion irradiation, may release an additional amount of the trapped carriers in the channel which in turn would raise the mobility of the electron gas. Ion implanted structures may consequently exhibit better output device characteristics by virtue of getting damaged to an extent where they form defects which help the interface relax in such a manner that while undergoing the reversal effect (annealing to recover the ion-induced damage) they provide more carriers to the channel with greater mobility and lower sheet resistivity. This way, relatively higher initial carrier density materials may also be used to perform device operations exactly the same way as could be done with lower initial carrier densities in AlGaN/GaN structures. Therefore, our suggested ion implantation model and designed experiments provide a substantial control on the carrier transport mechanisms while looking at the output device characteristics of AlGaN/GaN HEMT structures. This discussion is exhibited in Figure5, where initial sheet carrier concentration is plotted against the 2DEG hall mobility in AlGaN/GaN layers for HEMT structures with and without implantation. The samples which were irradiated with combined implantation schedule of Helium and Silicon ions, received the fluence which was at least two orders of magnitude below the threshold doses that were used in earlier experiments for each independent case of initial sheet carrier concentration. This is due to the fact to produce light damage in the lattice which is neither stable nor long range to compensate the layer fully. This is an extent of ion-induced damage which is just about sufficient to trap the carriers at interface and recoverable at a low thermal budget. Figure 6 clearly shows that the higher values of 2DEG mobility are achieved when the HEMT structures have undergone a carefully designed ion implantation and annealing at temperatures around 400°C. This trend is repeatable for starting materials with different initial carrier densities.

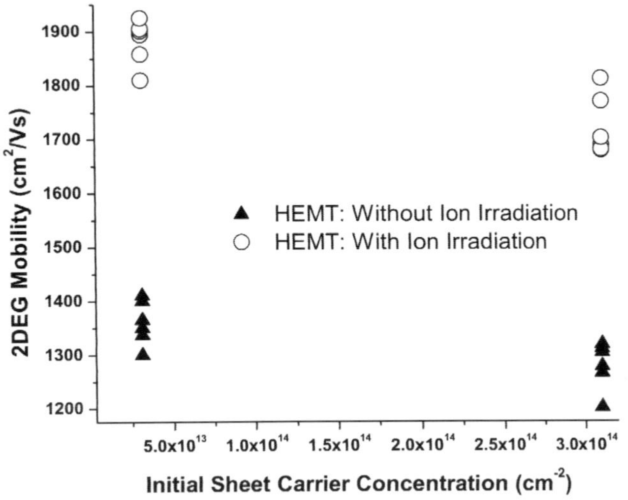

Figure 5. 2DEG Hall mobility versus initial channel carrier density (n_s= 3.1x10^{13} cm^{-2} and n_s= 3.1x10^{14} cm^{-2}) for AlGaN/GaN structures with and without ion implantation and annealing cycles (combined He^+ and Si^+ implants with ion doses 2 orders of magnitude lesser than the Threshold Dose in each case as shown in Fig.2)

A converse is plotted in Figure 6 where sheet resistivity is plotted against the starting carrier density of AlGaN/GaN HEMT structure for cases where the structures are measured with and without the ion irradiation received. A converse pattern is achieved as compared to Figure 5. It is interesting to note that the lattice engineering due to the carefully designed ion implantation schedule and subsequent rapid thermal annealing provided a very low value of sheet resistivity (~100 ohms/square) which is better than many commercial devices manufactured by conventional process routines without introducing lattice normalization engineering by ion implantation.

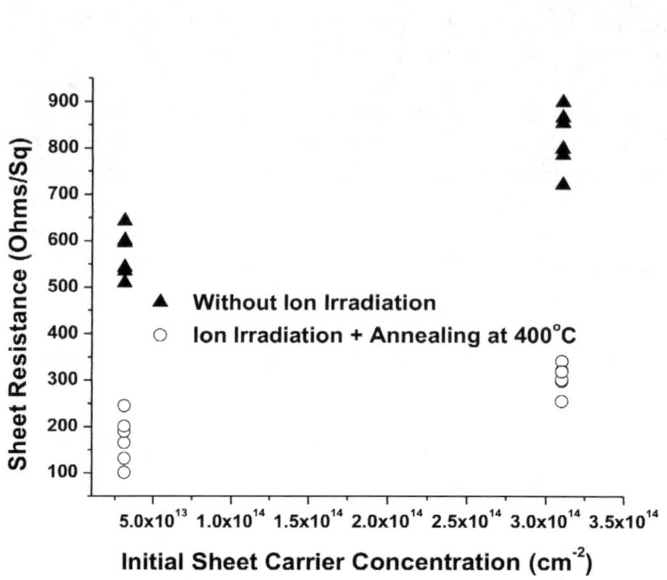

Figure 6. Sheet Resistivity versus initial channel carrier density (n_s= 3.1x10^{13} cm^{-2} and n_s= 3.1x10^{14} cm^{-2}) for AlGaN/GaN structures with and without ion implantation and annealing cycles (combined He^+ and Si^+ implants with ion doses 2 orders of magnitude lesser than the Threshold Dose in each case as shown in Figure 2)

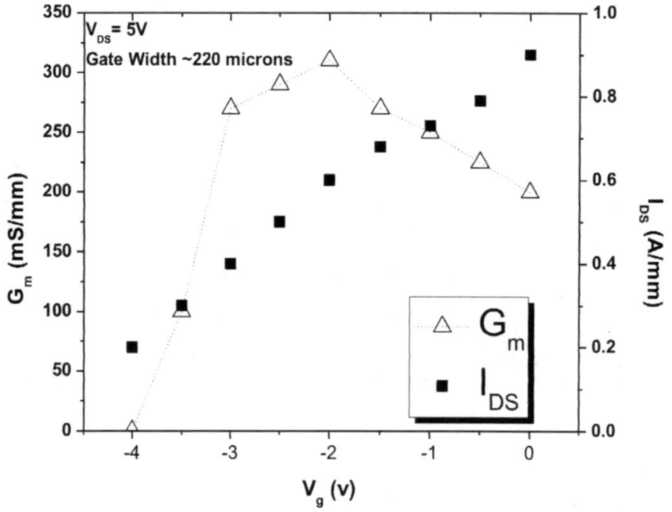

Figure 7. Characteristic trends of extrinsic transconductance (G_m) and drain current density (I_{DS}) versus the gate voltage (V_g)

Both the lower values of sheet resistivity and higher value of 2DEG mobility are vital to improve the RF characteristics of the HEMT device for small signal applications. Figure 7 exhibits the characteristic trends of extrinsic transconductance (G_m) and drain current density (I_{DS}) versus the gate voltage (V_g). The measurements were performed with the V_{DS} = 5 Volts supplied to the structure having gate dimensions of ~0.4x220 μm^2. A maximum value transconductance of over 300 mS/mm was achieved for ion-implanted structures, which is better than the average commercial HEMT structure without undergoing the post process ion implantation and annealing cycles.

4. Conclusion

The ion implanted AlGaN/GaN High Electron Mobility Transistors (HEMT) test structures were studied thoroughly to look into the possibilities of enhancing the HEMT efficiency for high-power and high-frequency electronic, photonic and gas sensing applications. With the results of this detailed study, we conclude the following:

- A dedicated experimental design was created in order to study the influence of the physical parameters in response to the high energy (by virtue of in-situ beam heating due to highly energetic implantation) ion implantation to the active device regions in nitride HEMT structures. Several sets of simulations were performed to predict the appropriate experimental conditions during the process of ion implantation induced damage formation before the test structure actually undergone the ion irradiation. A very calculated disorder or damage created in the HEMT structure was then studied carefully with Electrical characterization techniques such as Hall, I-V and G-V measurements.
- The evolution of the electrical characteristics affecting the high-power and high-frequency operations was also analyzed by subjecting the HEMT active device regions to the progressive time-temperature annealing cycles.
- The influence of relatively damaged HEMT structure by using "carefully" designed ion irradiation protocols was compared with the ones which are commercially available (having no post-process implantation protocols experimented) and it revealed that by changing the ion-induced physical dynamics of the channel and interface regions in AlGaN/GaN HEMT structures may enhance the operational efficiency of the device in many ways.
- The suggested model can also provide a functional process window to control the extent of 2D Electron mobility in the AlGaN/GaN HEMT devices undergoing a full cycle of thermal impact i-e from a desirable conductive region to a highly compensated one. That is shown possible by engineering the HEMT structures with or without the irradiation of so-called Threshold Doses ($\boldsymbol{D_{th}}$) and their consequent impact on the carrier density of the channel in the AlGaN/GaN HEMT devices. The evolution of such a physical impact is also studied by varying the annealing temperatures to devise process instruments for various operational domains in AlGaN/GaN HEMT structures. These results and analysis have ramifications for HEMT device design and fabrication process protocols in the manufacturing of opto-telecommunication components and ultra-high efficiency gas sensors.

References

[1] Mimura T 2002 *IEEE J Microwave Theory and Tech* **50** 780
[2] Ziegler J F 2010 *Stopping and Range of Ions in Matter (SRIM)* (USA: Org.Annapolis MD)
[3] Roman W 2008 *NNIN REU Research Accomplishments* Washington DC pp 164 – 165
[4] Kryzhanovskaya N V et al 2005 *Semicond. Sci. Technol. Bimberg* **20** 961
[5] Kurtz S R, Modine N A, Jones E D, Allerman A A, and Klem J F 2002 *Semicond Sci Technol.* **17** 843
[6] Ahmed S, Too P, Gwilliam R, and Sealy B J 2001 *Appl Phy Lett* **79** 3533
[7] Ahmed S, Amirov K, Larsson U, Lin J, Haq A, Too P, and Tabatabaian Z 2005 *Vacuum* **78** 137
[8] Ahmed S, Lin J, Haq A, and Sealy B 2005 *Nucl Instrum Meth Phys* B 240 214
[9] Ahmed S, Lin J, Haq A, Sealy B and Gwilliam R 2005 *Nucl Instrm Meth Phys B* **237** 102
[10] Ahmed S, Gwilliam R, and Sealy B J 2001 *Semicond. Sci Technol* **16** 28

[11] Ahmed S, Knights A P, Gwilliam R and Sealy B J 2001 *Semicond Sc. Technol* **16** 17
[12] Ahmed S, Gwilliam R and Sealy B J 2001 *Semicond. Sci. Technol* **16** 64
[13] Ahmed S, Nawaz R, Taiq R, Amirov K, Larsson U and Syed W A 2007 *Appl Phys Lett* **91** 062112
[14] Boudinov H, Coelho A V and DeSouza J P 2002 *J Appl Phys* **91** 6585
[15] Pearton S J 1990 *Mater Sci Rep* **4** 313
[16] Danilov I, DeSouza J P and Boudinov H 2002 *J Appl Phys* **92** 4261
[17] Ahmed S, Sealy B J, and Gwilliam R 2002 *IEEE Technical Digest: Electron Devices for Microwave and Optoelectronic Applications*, 18
[18] DeSouza J P, Danilov I, and Boudinov H 1997 *J Appl Phys* **81** 650
[19] De Souza J P, Danilov I and Boudinov H *Appl Phys Lett* **68** 535
[20] Ahmed S, Amirov K, Larsson U, Too P, Sealy B J and Gwilliam R 2003 *IEEE Technical Digest: Electron Devices for Microwave and Optoelectronic Applications* 210

ICSICCST 2013 IOP Publishing
IOP Conf. Series: Materials Science and Engineering **51** (2013) 012009 doi:10.1088/1757-899X/51/1/012009

Bio-medical imaging: Localization of main structures in retinal fundus images

A Basit[1] **and S J Egerton**
School of Information Technology, Monash University, Sunway Campus, Jalan Lagoon Selatan, Bandar Sunway, 46150, Selangor, Malaysia

e-mail: abdulbasit1975@gmail.com, simon.egerton@monash.edu

Abstract. Retinal fundus images have three main structures, the optic disk, fovea and blood vessels. By examining fundus images, an ophthalmologist can diagnose various clinical disorders of the eye and the body, typically indicated by changes in the diameter, area, branching angles and tortuosity of the three main retinal structures.Knowledge of the optic disk position is an important diagnostic index for many diseases related to the retina. In this paper,localization of optic disc is discussed. Optic disk detection is based on morphological operationsand smoothing filters.Blood vessels are extracted using the green component of a colour retinal image with the help of a median filter. Maximum intensity values are validated with blood vessels to localize the optic disk location. The proposed method has shown significant improvements in results.

Introduction

Early diagnosis of retinal diseases can prevent vision loss in large populations. Currently, Diabetic retinopathy, glaucoma and agerelated macular degeneration are the main diseases which can lead to blindness if not treated earlier. To detect changes in retina, image processing techniques are being developed to facilitate and share the burden of ophthalmologists in screening programs.

Many automatic methodshave been developedfor the localization of retinal structures (pl. see **Figure 1**)duringthe last decadebecause manual localization is a very time consuming and tedious job which requireshighlevels of concentration,a lot of training. It takes time for an ophthalmologist to become proficiently skilled. In normal retinal fundus images, the Optic Disk (OD) is a circular shaped bright yellow structure. Blood vessels enter the eye through the OD. Location of the OD and its diameter is also important because its examination discloses information about certain diseases of the eye.Its diameter is used as a reference for detecting the macula region in the retinal image. Many methods have been proposed to find the OD location including general shape based approaches like the Hough

[1] Permanent Address: TPPD, PINSTECH, Nilore, Islamabad, Pakistan

transform [1], template matching [2], blood vessels convergence [3], variance of intensity [4], modified active contour [5]and so on.

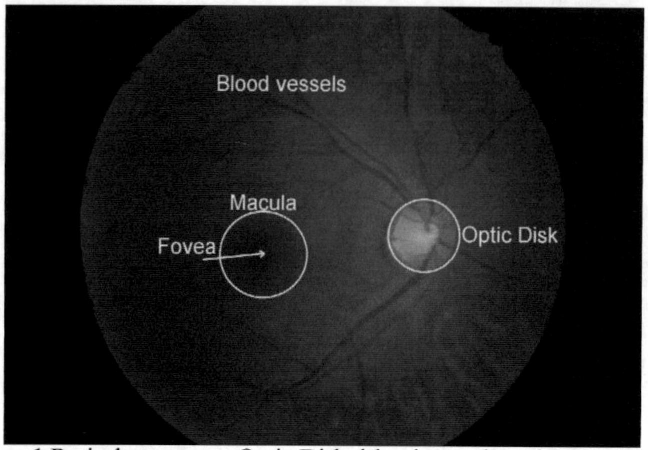

Figure 1. Retinal structures: Optic Disk, blood vessels and Fovea/Macula

In this paper, an OD localization method is proposed based on two facts, firstly that the OD is located at the convergence ofthe main blood vesselsand secondly it is the brightest part of the image. The method first segments the fundus-image to identify the main blood vessels and then uses the resultant blood vessel map to efficiently identify the brightestregion and with it the OD centre.

Methodology

In retinal images the OD is the bright yellow part which lies on the widest (main) blood vessel.Steps involved in the proposed method are shown in Figure 2.

A colour fundus retinal-image is used as the input to the method, which automatically outputs the location of the OD as the result. Each colour image is processed using the raw Red, Green and Blue (i.e. RGB) colour space.It is observed that in the Green component the blood vessels are very prominent as compared to the other two components whereas in OD is clearer in the Red colour channel. So in order to find the main blood vessels the Green component is extracted to form a vessel map, this map is then smoothedusing a median filter, this produces a final diffused vessel map. The vessel map is used to mask the original image, isolating the actual blood vessels to produce an actual blood vessel map. This actual blood vessel map is then used to validate the coordinates of the point selected as the location of maximum intensity. To find the location of maximum intensity the green component is first smoothed with a small median filter, large enough to usefully reduce image noise, but small enough to retain fine edge information. Typically the smoothed image will have many pixels with the same intensity, so in order to locate the OD, a median of these maximum intensity points is calculated along the x and y axes and the closest integer coordinate is selected as the candidate centre point for the OD. The candidate point is then validated against the actual blood vessel map if the validation fails then the next best candidate point is considered. This process continues till the exact localizing point in OD is obtained. It is important to note that if this search could not find any point with the described criteria then first encountered point is considered the localization point.

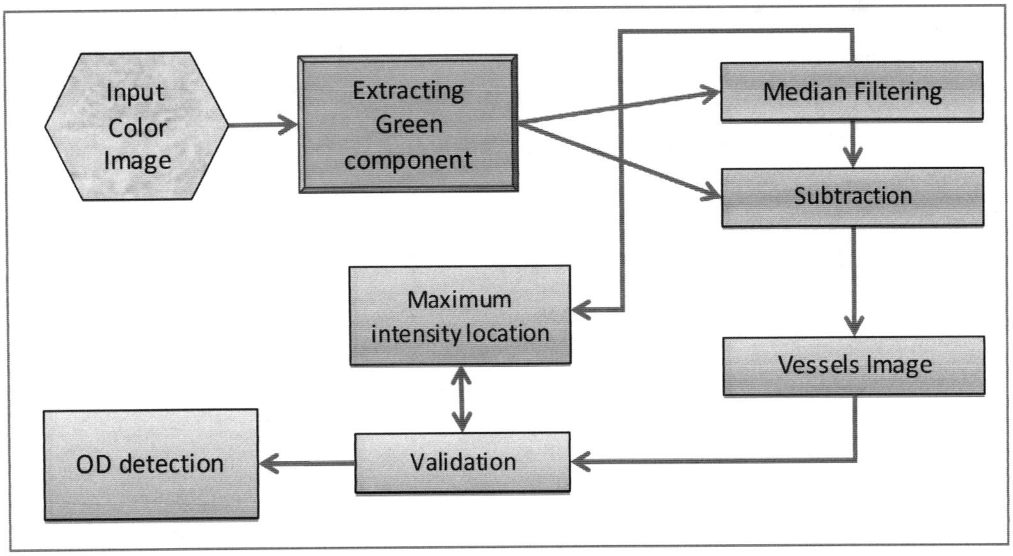

Figure 2. Optic disk detection steps

Most of the times, program outputs a valid point whereas on a few pathological images program fails to find the actual location in the OD because in these image vessels image is not complete which diverts to incorrect localization. In future, a better vessel image extractor will be incorporate to overcome this problem.

Results and Discussion

In this paper, experiments on to publically available retinal databases are reported. DRIVE[6], the mostly commonly used database in this research domain, contains 40 colour images of the retina. These images are captured by a Canon CR5 3CCDcamera with a 45 degree field of view (FOV), with 565×584 pixels and an 8 bit depth per RGB colour component. The database is divided into two sets,a test set and a training set with 20 images each.Four images with pathologies are included in the test set.

The DIARETDB1 [7]database contains 89 colour retinal fundus images. Eight four images have at least mild non-proliferative signs of diabetic retinopathy, and 5 are normal images with no signs of the diabetic retinopathy according to all experts who took part in an evaluation of the dataset.Allimages are captured with a 50 degree FOV usinga digital fundus camera with varying settings.This dataset is referred to as "calibration level 1 fundus images".

The OD is manually segmented by experts to evaluate the efficiency of algorithmic results. If the detected point is within the manually segmented image of the OD thenit is considered a correct localization of the OD. Experimental results on the mentioned databases are given in Table 1. In the DRIVE database, the OD is localized in all the images successfully. In DIARETDB1 database, seven out of 89 images are not correctly localized by the proposed method. The proposed method used both intensity and blood vessels in consideration when finding the OD location. The algorithm performs excellent on DRIVE database because it does not contain severely pathological images which are present in DIARETDB1.

Table 1. Results of optic disk localization

Database	**No of images**	**Success Rate**
DRIVE	40	100%
DIARETDB1	89	92.13%

Results are compared with other methods in Table 2.

Table 2. Comparison of results with other methods[a]

	Success Rate (%)	
METHOD	**DRIVE**	**DIARETDB1**
Sopharak et al. [8]	95	59.55
Walter et al. [9]	77.5	92.13
Stapor et al. [10]	85.5	78.65
Seo et al. [11]	95	80.89
Lupascu et al. [12]	95	86.51
Kande et al. [13]	95	88.76
Pereira et al. [14]	100	93.25
Malek et al. [15]	100	89.00
Proposed method	100	92.13

[a] Results of other methods are obtain from the literature

Conclusion

Retinal fundus images a main source of diagnosis for retinal diseases. The main structures present in the retinal images are blood vessels, macula/fovea and the optic disk.Knowledge of the OD position is an important diagnostic index for many diseases. In this research work, a method is proposed to localize the OD from the retinal images based on maximum intensity locations and blood vessels. Experiments are performed on two databases which are publicly available. Proposed method shows promising results. As future work, we will includesegmentation and boundary information of the OD andimprove the methods by considering other properties related to the OD.

References

[1] S. Sekhar, W. Al-Nuaimy, and A. K. Nandi, "Automated localisation of retinal optic disk using Hough transform," in Biomedical Imaging: From Nano to Macro, 2008. ISBI 2008. 5th IEEE International Symposium on, 2008, pp. 1577-1580.
[2] M. Lalonde, M. Beaulieu, and L. Gagnon, "Fast and robust optic disc detection using pyramidal decomposition and Hausdorff-based template matching," Medical Imaging, IEEE Transactions on, vol. 20, pp. 1193-1200, 2001.
[3] A. Hoover and M. Goldbaum, "Locating the optic nerve in a retinal image using the fuzzy convergence of the blood vessels," Medical Imaging, IEEE Transactions on, vol. 22, pp. 951-958, 2003.
[4] C. Sinthanayothin, J. F. Boyce, H. L. Cook, and T. H.Williamson, "Automated localization of the optic disc, fovea, and retinal blood vessels from digital color fundus images," British Journal of Ophthalmology, vol. 83, pp. 902-910, 1999.
[5] J. Xu, O. Chutatape, and P. Chew, "Automated optic disk boundary detection by modified active contour model," Biomedical Engineering, IEEE Transactions on, vol. 54, pp. 473-482, 2007.
[6] "DRIVE: Digital Retinal Images for Vessel Extraction, http://www.isi.uu.nl/Research/Databases/DRIVE/," 2004.
[7] "DIARETDB1, http://www2.it.lut.fi/project/imageret/diaretdb1/ accessed," 2010.
[8] A. Sopharak, B. Uyyanonvara, S. Barmanb, and T. H. Willliamson, "Automatic detection of diabetic retinopathy exudates from non dilated retinal images using mathematical morphology methods," Computerized medical Imaging and Graphics, vol. 32, pp. 720-727, 2008.
[9] T. Walter and J. C. Klein, "Segmentation of color fundus images of the human retina: Detection of the optic disk and the vascular tree using morphological techniques " Proceedings of the 2nd International Symposium on Medical Data Analysis, pp. 282-287, 2001.
[10] K. Stapor, A. Switonski, R.Charastek, and G.Michelson, "Segmentation of eye fundus images using methods of mathematical morphology for glucoma diagnosis," ICCS 2004 ,4th international conference,Krakow,poland,june 6-9,2004,Proceedings,part IV ,lecture notes in computer science, pp. 41-48, 2004.
[11] J. M. Seo, K. K. Kim, J. H. Kim, K. S.Park, and H.Chung, "Measurement of ocular torsion using digital fundus image," Proceeding of the 26th Anual International Conference of the IEEE EMBS,San francisco,USA,, pp. 1711-1713, 2004.
[12] C. A. Lupascu, D. Tegolo, and L. D. Rosa, "Automated detection of optic disk location in retinal images," 21st Iee international symposium on computer based medical systems,IEEE,university ofJyvaskyla,Finland, pp. 17-22, 2008.
[13] G. B. Kande, P. Subbaian, and T. Savitri, "Segmentation of Exudates and Optic disk in retinal images," Sixth indian Conference on computer Vision,Graphics and Image processing,2008,ICVGIP,IEEE Computer Society.Bhubaneswar,India, pp. 535-542, 2008.

IOP Conf. Series: Materials Science and Engineering **51** (2013) 012009 doi:10.1088/1757-899X/51/1/012009

[14] C. Pereira, L. Gonçalves and M. Ferreira, "Optic disc detection in color fundus images using ant colony optimization." Medical & biological engineering & computing (2013): 1-9.
[15] J. Malek, M. B. Abdallah, A. Mansour and R. Tourki. "Automated optic disc detection in retinal images by applying region-based active aontour model in a variational level set formulation." In Computer Vision in Remote Sensing (CVRS), 2012 International Conference on, pp. 39-44. IEEE, 2012.

IOP Conf. Series: Materials Science and Engineering **51** (2013) 012010 doi:10.1088/1757-899X/51/1/012010

THE EMERGING HIGH EFFICIENCY VIDEO CODING STANDARD (HEVC)

Gulistan Raja[1] **and Awais Khan**
Department of Electrical Engineering,
University of Engineering and Technology, Taxila

E-mail:gulistan.raja@uettaxila.edu.pk

Abstract. High definition video (HDV) is becoming popular day by day. This paper describes the performance analysis of latest upcoming video standard known as High Efficiency Video Coding (HEVC). HEVC is designed to fulfil all the requirements for future high definition videos. In this paper, three configurations (intra only, low delay and random access) of HEVC are analyzed using various 480p, 720p and 1080p high definition test video sequences. Simulation results show the superior objective and subjective quality of HEVC.

1. Introduction

Imagine that you want to transmit or store a TV quality digital video. Transmission or storage capability of 37.32 Megabytes is required for 1 second of video and 134. 37 Giga bytes are required for 1-hour uncompressed (raw) video program. This requires enormously high data transmission and/or storage medium, which is beyond the capabilities of today's systems. Therefore, there is a need for compression to deal with this kind of high-bit rate data. Moreover, increasing demand for superior level quality and higher resolution video calls for need of high definition video in comparison to standard video. Applications like Blue-ray, HD-DVD, HDTV, IPTV, digital cable, dish HD, Web HD downloads etc. are becoming common day by day. The amount of data to be processed and transmitted in case of HDV is significantly high and this requires video compression for efficient archival and transmission. The growth in video coding is mainly fuelled by two international organizations: Moving Picture Expert Group (MPEG) and International Telecommunication Union (ITU). MPEG is working group of experts from International Organization for Standardization (ISO) and International Electrotechnical Commission (IEC) [1-2]. The MPEG-1 and MPEG-4 visual standards are developed by MPEG while H.261 and H.263 standards are developed by ITU-T. ISO/IEC and ITU-T also worked jointly to produce H.262/MPEG-2 and H.264/MPEG-4 standards. The upcoming HEVC is joint effort of ISO/IEC MPEG and ITU-T VCEG known as Joint Collaborative Team on Video Coding (JCT-VC). Performance analysis of HEVC standard shows that it can achieve significant bit rate reductions in comparison to existing standards for equal perceptual quality video [3-4]. HEVC provides more than 50% bitrate saving than its successor video standard H.264 by providing the same subjective and objective video quality [5]. It provides more flexible

[1] To whom any correspondence should be addressed.

block structure, more intra prediction modes and wide range of transform unit (TU) and reduced the complexity of in-loop filter than the previous H.264 video standard [6].

2. Performance Analysis of High Efficiency Video Coding

The High Efficiency Video Coding (HEVC) is a latest video coding standard developed by ITU-T Video Coding Expert Group (VCEG) and ISO/IEC Moving Picture Experts Group (MPEG) in a Joint Collaborative Team on Video Coding (JCT-VC) [7]. The main HEVC incorporated features in comparison to its predecessor video coding standard H.264/AVC are [8]:

- Coding tree units and coding tree block structure
- Transform units and transform blocks
- Intrapicture prediction
- Enhanced entropy coding
- Modified loop filtering.

We have used three resolutions of 480p (832 X 480), 720p (1280 X 720) and 1080p (1920 X 1080) high definition video sequences for analysis [9]. All the sequences are in YUV format and 50 frames for each sequence at the rate of 25 frames per second.

Table 1 PSNR and bitrate comparison for various configurations of HEVC

Test Sequence	QP	Intra only		Low Delay		Random Access	
		Y-PSNR	**Bitrate**	**Y-PSNR**	**Bitrate**	**Y-PSNR**	**Bitrate**
Basketball Drive (1920 X 1080)	23	40.659	24472.901	39.481	7224.701	39.352	5421.24
	28	39.006	11284.320	37.961	2945.587	37.944	2200.39
	33	37.486	6253.603	36.098	1438.483	36.186	1099.84
	38	35.524	3537.797	34.036	773.515	34.242	605.65
	43	33.205	2027.918	31.726	407.035	32.033	392.318
	48	30.745	1030.421	29.314	204.278	29.680	198.590
Cactus (1920 X 1080)	23	39.860	55153.891	38.155	10261.690	38.054	8081.11
	28	37.402	27207.254	36.152	3487.598	36.377	2953.84
	33	35.059	14828.453	33.984	1620.432	34.393	1446.44
	38	32.525	7928.270	31.624	815.141	32.160	757.25
	43	29.946	4165.704	29.173	401.237	29.779	460.651
	48	27.551	1886.458	185.347	26.845	27.404	208.555
Four People (1280 X 720)	23	43.341	13586.122	42.115	1013.131	42.375	1173.912
	28	40.709	8367.614	39.889	444.158	40.363	579.960
	33	37.783	5148.691	37.262	235.507	37.875	327.259
	38	34.600	3088.517	34.316	132.178	35.016	192.562
	43	31.353	1730.890	31.209	71.424	31.903	108.005
	48	28.332	862.642	28.228	36.370	28.924	55.224
Kristen and Sara (1280 X 720)	23	44.026	9999.490	42.784	909.648	42.993	974.146
	28	41.614	6003.355	40.685	376.829	41.147	459.149
	33	38.888	3620.342	38.248	189.466	38.869	247.910
	38	35.913	2164.210	35.510	103.901	36.246	142.454
	43	32.865	1266.624	32.612	58.075	33.414	81.922
	48	29.935	680.011	30.442	29.753	30.519	44.866
Race Horses (832 X480)	23	41.442	15682.042	38.596	5663.299	37.707	4722.566
	28	37.765	9614.333	34.769	2353.046	34.452	2077.723
	33	33.975	5439.782	31.412	1051.603	31.557	1000.032
	38	30.297	2601.533	28.585	486.293	28.863	490.114

	43	27.516	1096.152	26.354	213.163	26.523	217.824
	48	25.810	528.298	24.639	98.016	24.827	100.464
Keiba (832 X480)	23	41.843	9424.771	39.617	3084.312	39.123	2928.566
	28	38.822	5451.624	36.526	1368.691	36.223	1400.136
	33	35.768	3071.006	33.506	648.902	33.386	687.605
	38	32.753	1647.686	30.823	332.563	30.803	368.693
	43	29.919	874.862	28.278	170.107	28.344	195.106
	48	27.127	428.323	25.807	84.245	25.883	98.299

The quantization parameter (QP) range is kept from 23 to 48 to determine the behaviours of the three configurations. Table 1 describes the comparison of bit rate and Peak to Signal Noise Ratio (PSNR) for intra only, low delay and random access configurations of HEVC. It is observed from table 1 that intra only configuration performs better in terms of PSNR in comparison with other two configurations. However, bitrate requirement for intra only is higher with respect to other two configurations due to the fact that prediction is not used in intra only. In terms of bitrate saving, random access configuration of HEVC performs better at cost of PSNR.

Keiba (832 X480), Frame 28

(a) (b) (c)

Four People (1280 X720), Frame 18

(a) (b) (c)

Basketball Drive (1920 X 1080), Frame 31

(a) (b) (c)

Figure 1. Subjective comparison of 480p, 720p and 1080p high definition sequences for various configurations (a) intra only (b) low delay (c) random access

Figure 1 shows the subjection comparison of HEVC's intra only, low delay and random access configurations for Keiba, Four people and Basketball drive high definition test sequences. It is observed that perceptual quality seems to be same but close observation reveals the better quality of intra only configuration in comparison to others.

3. Conclusion

We have performed performance analysis of all the encoder configurations used in HEVC standard by using various types of high definition video test sequences. Experimental results show that video quality of the sequences encoded by using intra only configuration is better than other two configurations of HEVC. Moreover, random access configuration provides the maximum bitrate saving among all configurations of HEVC without significantly affecting the subjective and objective quality of video.

4. References

[1] Sikora T 1999 *Signal Process. Multimedia* 225.
[2] Kalva H, Lee J-B 2012 Springer-Verlag Berlin Heidelberg 1.
[3] Sullivan G. J, Ohm J-R, Han W-J, Wiegand T 2012 IEEE Trans. Circuits Syst. Video Technol. 22 1649.
[4] Ohm J-R, Sullivan G J, Schwarz H, Tan T K, Wiegand T 2012 IEEE Trans. Circuits Syst. Video Technol. 22 1669.
[5] Raja G, Khan A, Rashid A, Khan AK 2013 Life Science Journal 10 800.
[6] Ugur K, Andersson K, Fuldseth A, Bjøntegaard G, Endresen LP, Lainema J, Hallapuro A, Ridge J, Rusanovskyy D, Zhang C, Norkin A, Priddle C, Rusert T, Samuelsson J, Sjoberg R, Wu Z 2012 IEEE Transactions on Circuits and Systems for Video Technol 22 1688.
[7] ITU-T SG16 WP3 and ISO/IEC JTC1/SC29/WG11 2013 JCTVC-L1003 v34 14.
[8] Pourazad M T, Doutre C, Azimi M, Nasiopoulos P 2012 IEEE Consumer Electro. Mag. 1 36.
[9] High Definition Test sequences 2013 *ftp://ftp.tnt.uni-hannover.de/testsequences*.

IOP Conf. Series: Materials Science and Engineering **51** (2013) 012011 doi:10.1088/1757-899X/51/1/012011

Possible Benefits of Mycorrhizal Symbiosis, in Reducing CO2 from Environment

Rafia Azmat[1]
Department of Chemistry
University of Karachi, Karachi, Pakistan
Email:rafiasaeed200@yahoo.com

Abstract. It is a fact that the relationship between a fungus and a plant can have a great impact on the environment, especially under drought conditions. Experiments conducted at the laboratory scale suggested that in mycorrhizal symbiosis; plants usually provide their fungal partners with carbohydrates from photosynthesis and receive mineral nutrients. It is observed that mycorrhizal inoculated plants observed large surface area of leaves and outsized root sections which were helpful in increasing the rate of photosynthetic processes. This may be attributed to the rapid production of carbohydrate for their fungal mate. The same phenomena can be observed in environments of high traffic density or waste burning, industrial zones (where there are emissions of CO_2 from chimneys) or the areas that are lack nutrients such as nitrogen and phosphorus. It may be observed that the plants that have this association with mycorrhizal fungi may obligate a better chance in inhabiting this area. These plants can be helpful in reducing the CO_2 from the polluted atmosphere. The large length of the roots were related to the absorption of water molecules for survival as well as formation of first organic complex CHO for providing the energy to the plant in biotic stress and C and nutrient exchange between fungal partner and plants.

Key Words: plant, mycorrhiza, surface area, CO_2, CHO

Introduction

Carbon dioxide in Earth's atmosphere

Although CO_2concentration is relatively small in the atmosphere. It is vital component of Earth's atmosphere because it absorbs and emits infrared radiation at wavelengths of 4.26 µm (asymmetric stretching vibrational mode) and 14.99 µm (bending vibrational mode), thereby playing a role in the greenhouse effect. The concentration of CO_2in Atmospheredetermines its contribution to the greenhouse effect and the rates of plant and algal photosynthesis (lives process)[1-3]. In 21st century the concentration has increased markedly, at a rate of 2.0 ppm/yr during 2000–2009 and faster since last decades. It was 280 ppm in pre-industrial times, and has risen to 400 ppm (parts per million) as of May 2013, with the increase largely attributed to anthropogenic sources. It is also produced by various microorganisms from fermentation and cellular respiration. Plants, algae and cyanobacteria convert carbon dioxide to carbohydrates by a process called photosynthesis. They gain the energy needed for this reaction from absorption of sunlight by chlorophyll and other pigments. Oxygen, produced as a by-product of photosynthesis, is released into the atmosphere and subsequently used for respiration by heterotrophic organisms and other plants, forming a cycle. Natural sources of atmospheric carbon dioxide include volcanic outgassing, the combustion of organic matter, wildfires and the respiration processes of

living aerobic organisms. Man-made sources of carbon dioxide include the burning of fossil fuels for heating, power generation and transport, as well as some industrial processes such as cement making.[4,5]

The present level is higher than at any time during the last 800,000 years and likely higher than in the past 20 million year[1-5]

Todays the leading Countries like USA, Australia and Taiwan focusing the capturing and reusing of CO_2which may aid coal, oil. Reducing CO_2 through funding on Research Project, CO_2 Capture for Algae Biofuel in Australia, advancement in carbon capture plant Taiwan inaugurates advanced carbon capture plant

VAM Fungi reducing CO_2Naturally

VAM fungi inhabitant the roots of host plants and execute absorption services for the plant growth. Several investigations have established that plants associated with VAM fungi show increased uptake of various constituents from the soil, including water, and macro and micronutrients compared to non-VAM plants. As a result, VAM fungi develop their host plants' ability to grow under conditions of scarcity stress or in mineral deficient soils. It is valuable in anassessable degree of drought stress for plants grown under droughty conditions where irrigation is not available. Fields without irrigation can produce increased crop yields by applying mycorrhizal fungi inoculants, or by stimulating root colonization of native VAM fungi. In sustainable atmosphere arbuscularmycorrhizas can play a key role in reducing CO2 from the heavy traffic areas where burning of fuel produced large amount of CO_2 .Mycorrhizal formation and its application on green belt plants, may provide an adaptive strategy which provides the plant with an augmented capability for nutrient capture and cycling in soils with low nutrient availability and an increased tolerance to environmental stresses. The experimental results suggest that significance of maintaining mycorrhizal variety and a functioning soil mycelium is vital to sustaining plant growth. It helps in growing plant with large surface area and long roots which were help full in continuous supply of C through sugar contents of the plant to the VAM fungus for the development of hyphae[6-8].

Action of VAM Fungus

1. VAM fungi expand the roots by adding their own expansive network of absorbing strands to mine the soil for water and the dissolved minerals carried therein[9].
2. VAM fungi affect the opening or closure of the breathing pores in leaves. These pores are called "stomates." Under conditions of drought stress, the plant will close the stomates to reduce water loss. VAM fungi can affect the closure of these pores and help provide more efficient water conservation[8-10].
3. The opening of stomata is also leads for the exchange of two gases[7].
4. The larger surface area of VAM plants is beneficial for removal excess of CO_2 and release of O_2[7,11].
5. VAM fungi raise water pressure (turgor) in plant tissue, thereby avoiding or suspendingdrooping. This supports cell function, allowing growth and photosynthesis to continue[8,13].
6. The larger the surface area related with the process of photosynthetic process and associated with the production of sugar contents[7-9](Fig 1&2)
7. These sugars contents are transported to the entire plants and also C for building hyphae of VAM[7,]
8. Mycorrhizae helps in increasing the water absorbing capacity of the root, moderate the leaf water prospective and enduring drooping point below scarcity anxiety, raise the stomatal movement[7,14].

9. In addition when plant agonized by antagonisticdeficiency conditions the VAM can encouragea following sequences of self-protective (a), the strengthening of cell wall of the plant (b), the addition of pathogenesis associated protein and (c) the support of secondary metabolites [15-18-22].

Figure 1 Non VAM Plant

Figure 2 VAM Plant

Conclusion

It is suggested that VAM inoculation in Trees growing in high traffic zone may be helpful in reducing CO_2 from those areas and these areas may be protected from the harm of the C emission. These plants may be helpful in releasing of O_2 through exchange of both gases to make environment of urban areas green. It is recommended to conduct more research on this project for the safety of environment from excesses of CO_2.

References

[1] Tans, Pieter. "Trends in Carbon Dioxide".NOAA/ESRL.Retrieved 2009-12-11.

[2] Carbon Budget 2009 Highlights, globalcarbonproject.org, retrieved 2012-11-02

[3] "Carbon dioxide passes symbolic mark". BBC. 10 May 2013. Retrieved 10 May 2013.

[4] Etheridge, D. M.; L. P. Steele, R. L. Langenfelds, R. J. Francey, J.-M. Barnola, V. I. Morgan (1996). "Natural and anthropogenic changes in atmospheric CO2 over the last 1000 years from air in Antarctic ice and firn". Journal of Geophysical Research 101 (D2): 4115–4128. Bibcode:1996JGR...101.4115E. doi:10.1029/95JD03410. ISSN 0148-0227.

[5] "IPCC Fourth Assessment Report: Climate Change 2007 (AR4)". Intergovernmental Panel on Climate Change.Retrieved 2013-05-12.[5] Sinha S, Singh S and Mallick S 2008 Comparative growth response of two varieties of Vignaradiata L. (var. PDM 54 and var. NM 1) grown on different tannery sludge applications: effects of treated wastewater and ground water used for irrigation. Envirn.Geochem.& Health 30 pp 407–422

[6] Anne, M. C. ,Appollonia, S , Pamela, H., Catherine, I. , Christy, W. , Anne-Laure, M. & Sullivan P. O. Construction of an Environmentally Sustainable Development on a Modified Coastal Sand Mined and Landfill Site—Part 2. Re-Establishing the Natural Ecosystems on the Reconstructed Beach Dunes. Sustainability 2(3), 717-741.2010

[7] Azmat R Possible Benefitsof Mycorrhizal Symbiosis, in Reducing CO_2 from EnvironmentMaterials Science and Engineering **(MSE) (Press)**

[8] AzmatR , Pervez, A KanwalUAn Important Role of Caretnoids in Protection of Photosynthetic Apparatus under VAM Inoculation on Momordicacharantia" Current Pharmaceutical Biotechnology' (press).

[9] Azmat R. A Plausible Mechanism of Biosorption in Dual Symbioses by Vesicular-ArbuscularMycorrhizal in *Momordicacharantia*(submitted*)*

[10] Barea J.M. ,J. , Palenzuela ,P. Cornejo, I. Sánchez-Castro, C. Navarro-Fernández, A. Lopéz-García, B. Estrada, R. Azcón, N. Ferrol, C. Azcón-Aguilar. Ecological and functional roles of mycorrhizas in semi-arid ecosystems of Southeast Spain.Journal of Arid Environments .75, 12, 1292–1301 .**2011**

[11] BinetM.N. , Sage, L. , Malan C. , Clément , J.C. , Redecker ,D. , D. Wipf, R.A. Geremia, S. Lavorel, &Mouhamadou B. . Effects of mowing on fungal endophytes an .arbuscularmycorrhizal fungi in subalpine grasslands. Fungal Ecology ., 4, 248–255. **2013**

[12] Erik V., Wilfred F. M. RölingHannes A. Gamper George A. Kowalchuk Herman A. Verhoef, &Marcel G. A. van der Heijden. Positive effects of organic farming on below-ground mutualists: large-scale comparison of mycorrhizal fungal communities in agricultural soils. ine: 22 **2010**

[13] Fitter, A.H. , T. Helgason, &A. Hodge. Nutritional exchanges in the arbuscularmycorrhizal symbiosis: Implications for sustainable agriculture. Fungal Biology .25, (1): 68–72. **2011**

[14] Gianinazzi S, Gollotte A, Marie-Noëlle B., Diederik,T. R.&Wipf .D. The key role of arbuscularmycorrhizas in ecosystem services.Agroecology. 20, (8): 519-530. **2010**

[15] Jeffries P. &Barea . J. M. ArbuscularMycorrhiza: A Key Component of Sustainable Plant–Soil Ecosystems. Fungal Associations, 9, 51-75 **2012**

[16] KlironomosJ.,Zobel, M, Tibbett M , Stock W. D., Rillig M. C, Parrent J. L., Moora .M, Koch A. M. , Facelli J. Facelli, M. , Dickie E., & J. D. Bever . Forces that structure plant communities: quantifying the importance of the mycorrhizal symbiosis .New Phytologist 189, 2, 366–370, **2011**

[17] Johnson D. , Martin F.,Cairney J. W. G. & Anderson I. C. The importance of individuals: intraspecific diversity of mycorrhizal plants and fungi in ecosystems. New Phytologist. . 194, (3), 614–628,.**2012**

[18] Johnson N. C. , Angelard C., Sanders I. R. &Kiers .E. T. Predicting community and ecosystem outcomes of mycorrhizal responses to global change.Ecology Letters. 16, 1, 140–153, **2013**

[19] Kupper H, Setlik I, Spiller F C and Prasil O 2002 Heavy metal-induced inhibition of photosynthesis: targets of in vivo heavy metal chlorophyll formation Journal of Phycology 38 pp 429–441

[20] Marcel G. A. van der Heijden. Mycorrhizal fungi reduce nutrient loss from model grassland ecosystems. Ecology. 91:1163–1171. **2010**

[21] Opik M., ,M. Metsis, T. J. Daniell, M. Zobel, & M. Moora. Large-scale parallel 454 sequencing reveals host ecological group specificity of arbuscularmycorrhizal fungi in a boreonemoral forest .New Phytologist. 184, (2), 424–437. **2009**

[22] Opik M. , Vanatoa A., Vanatoa E. , Moora M., Davison J., J. M. Kalwij, Ü. Reier, &M. Zobel· The online database Maarj*AM* reveals global and ecosystemic distribution patterns in arbuscularmycorrhizal fungi (Glomeromycota). New Phytologist. 188, (1), 223–241, **2010**

Biosensors a promising future in measurements

Muhammad Saleem

Chairman, Dept. of Civil Engineering,
Jubail University College, Jubail Industrial City, Saudi Arabia

E-mail: i_am_saleem@yahoo.com

Abstract. A biosensor is an analytical device which can be used to convert the existence of a molecule or compound into a measurable and useful signal. Biosensors use stimulus to translate changes to recognisable signals and have great importance to society. Applications include diagnosis tools for diseases, security appliances, and other biomedical equipments. Biosensors can also be used in the detection of pathogens and other microbes in foodstuffs, drugs and processing industries. Enormous progress and advancement has been witnessed in this area. Research and development in micro level systems serves to interface biology with novel materials such as nanomaterial. Development of high speed and accurate electronic devices tfor use in medicine and energy storage (such as biofuel cells) is one of the target areas. This paper discusses the importance, use and current and future trend in the application of biosensors.

1. Introduction

Humans have sensors to sense and recognize the environment around us. In order to delegate some of responsibilities to equipments/instruments in sensing humans developed sensor. Sensors are the devices which can reacts to a perceptible action, such as heat, light, or pressure [1]. Sensor then responds to that action by generating a signal that can be deduced. In the instrumentation industry biosensors are among the fascinating areas of research. Biosensor generally, determines the concentration of various constituents and other parameters of biological interest and contributing in a broad spectrum of applications. Application of biosensors dated back to 1962 when Clark and Lyons developed a fast andmore precise biosensor for glucose measurement. Biological sensors are analytical devices that detect biochemical and physiological changes. Transducers are essential to convert the particular biological and chemical change into electrical data which can identify different biochemical components of a complex compound to isolate the desired biochemical compounds [2].

At the moment biosensors have expansive application in medicine, agriculture, environmental monitoring and the bioprocessing areas [3, 4]. During last two decades great advances in the design of sensor architectures, the integration of biological systems with monolithic silicon and optical technologies has been observed. The development of effective electron-transfer systems and the configuration of direct immunosensors recently got prominent appraisal from the users. Recent progress in these areas has already led to the introduction of new-generation biosensors into the competitive diagnostics market place [5].

Engineering advances during last decade made possible the advances in of electro-chemical DNA biosensors which have molecular investigativecompetences [6]. These sensors based on electro-chemical DNA providesnumerous advantages over molecular detection methods. This includes the capability to examinethe multifaceted body fluids, better sensitivity, suitability with micro-fabrication technology, having lowerenergy requirement, compact and compatible handy devices [5, 7].

It is reported that, electrochemical sensors based on continuous potentiometric or amperometric detection of enzyme reaction products demonstrated excellent results in the development of biosensors. This makes the electrochemical transducers economical, rapid to use and portable [8].

Advances in nanomaterial provoked the development of more sophisticated sensors in the field of measurement. Biosensors developed from nanomaterial generally utilize restrained bio receptor probe which is selective for the analyte molecules under investigation. Nanomaterial developed is more responsive to analytes and provides sensitive biological and chemical sensors which exhibit distinctive properties [9].

Among the recently developed biosensors is one which is being used for sensing cytosolic concentration of the analyte cAMP (cyclic adenosine monophosphate) on the cell membrane [10]. A comparable system is developed to investigate cellular responses to instinctive. These assays are being frequently used in the pharmaceutical and biotechnology research [11]. During last decade application of biosensors were at peak demand in various fields of animal science, animal husbandry, food industry and ecological control and monitoring [6].

The future of biosensor is promising in broad area of applications. One may see thediversity of biosensors which has been developed based on artificial recognition component this innovation allows the applicability of biosensors in a wider range of analytes and means of transduction this has encompassed the lower limits of selectivity and sensitivity [12]. Miniaturization in field instrumentation also permits biosensors to be constructed as arrays. In the recent future biosensors will offer a powerful tool which will be drastically mutable method to analytical approaches. Medical and pharmaceutical industries motivated the development of swift and reliable analysis of laboratory samples. Recently in the area of environmental monitoring, optimization of response duration, selectivity, permanence and economy are the prime targets of research priorities [13].

Diverse range of biosensors has been developed and other is under investigation. These sensors are based on antibodies, enzymes, artificial membranes, cells, and even entire animal tissues. These biosensors can be used in environmental assessment and monitoring which may lead to efficient and reliable measuring systems in the future [14].

Literature available reveled that innovative discoveries originating from microelectronics, bioelectronics and nanotechnologies are now altering the meaning of environmental samples analysis. These consist of chemical and biochemical analytics which will provide clinicians new theranostic tools in the coming era [15].

Present study discusses a brief review of the biological sensors, their application and recent improvements in addition to their present and future uses. Furthermore, the technical importance, techniques, performance, benefits, and limitations of different types of biosensors are potted to present the current and future applications of electrochemical and other biosensors especially in the field of environmental assessment and monitoring.

2. **Biosensor and Measurements**

The need for rapid, efficient and reliable instruments to monitor the environment has motivated the instrumentation industry to develop state of the art expertisewith better methodologies which may have the capability to investigate the conventional and emerging analytes rapidly, reliably and economically. Biosensors proved to be the most attractive analytical tools for better monitoring and to implement environmental legislations in these areas [16].

2.1. Biosensors Developments for Environmental Monitoring

Conventional methods which are being used present day in environmental applications having high precision and accuracy with lowestpossible detection limits, however these sensors are more costly, time consuming, and sometimes entail skilled personnel to operate [3]. In order to fulfill the requirement of environmental monitoring biosensors found to be more efficient and dependable analytical tools to provide rapid, reliable, and sensitive measurements with relatively more economical way; many of them aimed for in situ analysis. Biosensors can be utilized by industries as well as by regulatory authorities which may provide flexibility and broad usage of instrument for testing and screening of variety of environmental samples [17].

Generally, biosensors are divided into two broad classes i) based on the signal transduction and ii) based on the bio recognition method. In the transducing element class biosensors can be characterized as electro-chemical, bioluminescence, optical, piezoelectric, and thermal sensors [18].

Electrochemical biosensors are more pliable to create compact instruments with compatible instrumental sensitivity. These biosensors having ability to even operate in turbid media [19]. Electrochemical biosensors emerge as the most commonly used biosensors in monitoring and diagnosis tests in environmental and clinical analysis. Electrochemical biosensors can be classifyasconductometric, potetiometric, and amperometric biosensors [18].

Bioluminescence or bioanalytical biosensors are designed on the basis of utilizing certain enzymes to radiate photons as a by-product of these reactions. Bioluminescent biosensors is mainly used for toxicity and water quality monitoring [20].Optical biosensors are moresuitable for direct monitoring systems. Optical biosensors are designed to detect and emit the light as a result of a biological and/or chemical reaction [20].

Piezoelectric biosensors developed by coating the surface of the biosensor with a selectively binding biologically active substance [18,21]. Piezoelectric biosensors are favorite because they provide real-time output, simplicity of use, broader pH range, and lower cost[22].

Calorimetric biosensors detect various material of concerndue to heat evolved from biochemical reactions when the analyte reacts with aappropriatebiologically active material such as an enzyme [23].

A variety of components in the environment needs monitoring and detection. In the area of environmental quality monitoring biosensors are the successful technology which can provide swifter and cost competitive results.

2.2. Toxicity Genotoxicity and Mutagenicity

The determination of toxicity is very important in environmental monitoring because it can provide the actual picture of the extent of environmental pollution originated from aqueous, gaseous effluent and sediment [20].It is implicit that physical and chemical analysis alone cannot give information to evaluate the environmental risk associated with the contaminant [24]. However, the impact of such pollutants can be evaluated, through the use of some biological gears [4].

Due to the increase in variety of pollutants and their persistence researchers searched for rapid and reliable sensors which can detect at lower limit. At present among popular sensors enzymatic sensorsgetting more interest of many researchers. These sensors working onthe principle ofinhibiting an specific enzyme of interest. At present most entertained biosensors which are used for the determination of various environmental compounds are enzymatic biosensors. In general, majority of environmental biosensors have focused on variety of bacteria and their morphology while eukaryotic biosensors are sporadic; specially the use of mammalian cells. Thesecells areextramultifacetedsystem than bacterial cells and can give a more subtle response thana bacterial cell. Mammalian cells can also retort to the estrogenic effects of various chemicals [25].

Sometimes completemicroorganisms also used to determine the potential toxicity inwater or other environmental sample. Cell-sense,is one of theamperometricsensorswhichuses E.Coli cells for quick check of toxicity in the environment. It was suggested as one of the fast and more accurate toxicity evaluation methods in the toxicity assessment demonstration program operated by the United Kingdom Environmental Agency [26].

Sensors are developed for various areas of environmental toxicology, such as geno-toxicity and mutagenicity; they are reported as “biosensors for environmental stresses”. These sensors are generally operates by interacting with the concernedtoxicant with genetically engineered microorganisms or there nucleic acids and they are premeditated to react to applied toxicity [4]. Genotoxicity is mainly concomitant in various environmental materials, such as aromatic hydrocarbon, chlorophenols, PCBs and PAHs, and many other componentsidentified aspotential cancer-causing pollutants [20]. Use of green fluorescent having base of protein used to device biosensor for detecting activity of various genotoxicant {28].

2.3. *Endocrine-effect biosensors*

It is reported by many researchers that many contaminants in an environment can produce adversative health effects by interfering with endogenous hormone systems of a living body known as EDCs. These EDCs are a group of substances which cannot be demarcated by their chemical nature,they can be due to their biological influence. As the binding ability of the chemicals toward the natural estrogen receptors (ER) can be measured biosensors are developed. Utilizing natural estrogen receptor of humans, the SPR biosensor used to determine the estrogens and xenoestrogensin addition to investigate the binding capabilities of concernedcomponents [9, 18]. Furthermore, other than piezoelectric and electrochemical biosensors [35], optical biosensors also developed [29] which are based on estrogen receptors.

2.4. *Organic compounds*

Organic pollutants in the environments mainly consist of proteins, carbohydrates, fats and nucleic acids in an assortment of combinations. Following are the major areas of application of biosensors:

2.4.1. PCBs. Even though the production was restricted and even banned in some countries since long, it is still an environmental contaminant of the concern. A variety of biosensors have been designed with diverse configurations to monitor the PCBs in the environment [20, 30].

2.4.2. Petroleum products and polycyclic aromatic hydrocarbons (PAHs). These compounds are globally suspected as carcinogen and are of great concern for public health and safety. These compounds are principally formed during pyrolysis or incomplete combustion of organic materials. These contaminants generally of great concern due to their known and suspected carcinogenic effect along with their persistent in the environment. Lee and co-workers in the year 2005 described an integrated mini biosensor system which was based on green fluorescent protein-based Pseudomonas which can be utilized for continuous water toxicity monitoring [20]. The biosensor was developed for its ability to determine phenol, benzene, ethylbenzene, toluene, and various other compounds in aqueous phase solutions. Researchers designed the biosensors working on the basis of the capacity of a plasmid which can carry the toluene-benzene transcriptional activator [20]. Biosensors also developed to determine the naphthalenein environmental samples. These Amperometric biosensors were fabricated by utilizing Sphingomonasyanoikuyae B1 [31]. In order to monitor benzo(a)pyrene (BaP) in environmental samples, fluoroimmunosensor madeup of fiberoptic was developed in recent past. These sensors are extremely sensitive because they are based on laser excitation along with the optical detection system [29].

2.4.3. Dioxins. One of the potential contaminant of environmental pollution is Dioxins. They are polychlorinated compounds and released in the environment as a by-product during various chemical processes utilizing chlorine. EPA hasa great concern on Dioxins because of its known carcinogenic effects on human health [30], therefore they are in the lists of potential Endocrine Disrupting Compounds (EDCs) [31]. The SPR biosensor which was developed to monitorthe PCBs in the environment alsoutilizedfor the monitoring of the dioxin 2,3,7,8- TCDD successfully[32]. During recent past, the DRESSA biosensor areprepared in hepatomecells whose working principle isbinding the dioxins to an aryl hydrocarbon receptor [33].

2.4.4. Phenols. Major contributor of phenolic compounds are pulp and paper industry however, duringmanufacturing of medicines, colours, and antioxidants these compounds may be formed. Among the various phenolic compounds, chlorophenols are significant contributors of environmental contaminants because these compounds having greater toxicity and bioaccumulation tendency in the ecological system [20]. Parellada and colleagues in 1998 developed an amperometric biosensor, with tyrosinasebounded in a higrogel on a graphite electrode was [34]. This biosensor was reasonablyassociated with other available tecneques being utilizedto measure the phenolic compounds in various samples. An optical fibre sensor having reflectancefor p-aminophenol was developed which is based on immobilised bis-8-hydroxyquinoline [35].

2.4.5. Surfactants. Surfactants are the key ingredient of detergents among which anionic surfactants are the prevalent in the environment. Utilizing Pseudomonas rathonis T microorganism an amperometric biosensor was developed which can detect the anionic surfactants In an study reserchers claimed to achieved detection limit for sodium dodecyl sulfate (SDS) below 0.75 mg/L [14]. Researchers also investigated the sensitivity and selectivity of biosensors developed on bacterial strains who are capable of degrading the anionic surfactants. Researchers reported that by measuring the dissolved oxygen uptake whichcan change thecurrent in the oxygen electrode a biosensor can measure the surfactants up to the concentration of 0.25 μg/L [18].

2.4.6. Biocides/pesticides. Many researchers worked on the detection of bioxides and pesticides by utilizing inhibition property of acetyl cholinesterase (AChE) and colin oxidase. Such type of biosensors utilized to detect organo-phosphorous and carbamate pesticides [18, 24]. A fiber-optic biosensor for the detection of pesticides carbaryl and dichlorvoshas also been developed. A three-layer structure in which enzyme cholinesterase was immobilized on the outer layer developed which was in attached to an intermediate sol–gel layer which incorporated bromcresol purple, accumulated on an inner glass plate. The detection limit is reported as low as 5.2 μg/l [31].

2.4.7. Hormones. The extensive use of hormones in the environment with the evidence of impact in aquatic animals at concentrations below pictogram per liter, have warned the environmentalist to act accordingly. The potential hazardous impact of hormonesis generally in the aquatic life. This is at present a promising area of interest for ecological research[18, 36]. Servos and colleagues in 2005 detected female hormones such as 17beta-estradiol and estrone, in various municipal wastewater treatment plants in Canada during their study in significant concentrations [36].

Rodriguez-Mozaz and coworkers in 2005 developed biosensor to determine hormones. Estrone, and some other organic pollutants, was investigated with the help of opticalimmune-sensor in various water bodies in concentrations having range of ng/l. In recent past researchers were working on the development and improvement of sensors for the swift detection of metabolites and testosterone [18, 37].

2.4.8. Antibiotics. At present the extensiveproduction and use of antibiotics initiated the issue of antibiotic resistance which in fact can be transferred to humans and release to the environment through excreta. These antibiotics may also reach the meat and milk products if used on animals as well and may disrupt the food chain [14]. Scientist developed a biosensor known as BIACORE 3000 to study interactionsulfamethazine and furosemide [38, 39].

2.4.9. Toxins. Presence of toxins in the environment is of great concern for health and safety point of view. In order to assess and monitor bacterial toxins variety of sensors areprepared in the past for foodstuff and ecological control [30]. A fibre optic biosensor and impedance-based immunosensorwas developed to investigate staphylococcal enterotoxin B in various samples [18]. A fibreopticbased biosensor which uses evanescent wave of a conical optical fibre for signal perception was prepared to measure the Clostridium botulinum toxin. The biosensor is rapid and enables to detect the toxin up to 5 ng/mL within 1 min [41].

2.5. Microorganisms
Conventional practices to control diseases caused by pathogensare to control the suspected pathogenic microorganisms in food industry and environmental assessment. However, in addition to various other disadvantages lengthy and cumbersome detection methods made them unpopular in the medical and environmental monitoring area. Detection of disease causing viruses, bacteria, protozoa, and helminthes are generally less reliable, time-consuming, and uneconomical.

Researchers involved in health and environmental areas are interested in rapid and more accurate biosensors development. Salmonella enteriditis and Listeria monocytogenes are detected by utilizing SPR immobilizing on the gold cladded sensor. The detection limit was as low as concentrations below to 106 cell/ml [18]. Researchers reported the detection of Salmonella typhimurium in water in less than 2 minutes by using an immunosensorworking on the basis of acoustic wave principle [29].

2.6. Biochemical oxygen demand
The conventional BOD test such as wrinkle method has been used in environmental monitoring since many decades. This method has manyattractions to use such as it is being used all over the world as a universal standard technique, it has low analysis cost due to the use of simple and inexpensive equipments. However, this method has some limitation such as it consume at least 5 days to complete the and required skilled person. Furthermore, method is not appropriate for online monitoring. Rapid measurement of BOD is possible if one may use modern biosensor-based procedures. Several biosensors have been developed to measure BOD specially for higher strength waste effluents. In recent past, Chee et al. demonstrated an optical fibre biosensor to determine low concentration BOD such as for surface water samples. Instrument required only 15 minutes determining the BOD present in the sample [18]. Some biosensors based on the photocatalysisare also introduced recently [17]. These BOD biosensors are based on a novel microbial membrane [14].

2.7. Inorganic compounds
High toxicity of various inorganic pollutants makes their determination important for the environmental protection. Following are applications of biosensors for some important inorganic pollutants:

2.7.1. Metals. One of the frequentcontaminants in the environment is metals which actually forced the researchers to obtain sensors which are rapid, reliable and having very low detection limit. Recent development has been made in the improvement of sensors exploiting intact microbial cells to determineheavy metals of concern. Whole-cell biosensors are more advantageous due to their capability to interact only with the available fraction of metal ions. At the other hand, common methods are not having the ability to distinguish between fractions of metals that are available or not available to the biological systems [42]. Heavy metals are common cause of inhibiting the activity of enzymes, and utilization of this phenomenon for the determination of hazardous toxic elements provides many advantages, such as method is simple and having better detection sensitivity [18]. Biosensorsfor the determination of other heavy metals ions which based on urease inhibition are generally applied for the measurement of mercury [14, 20].

2.7.2. Inorganic phosphate. Presence of inorganic phosphate in surface waters may cause the eutrophication and environmental degradation. Determination of inorganic phosphate in various water bodies is important to monitor the extent of eutrophication [43]. Variety of enzymatic biosensors have been developed in past to determine phosphate level in the environment [18]. Parellada and coworkers in 1998 developed biosensor which works on the principle of sequential action of three enzymes that opens up a way to construct reagent less sensor [44]. Conrath and co-workers developed an enzyme sensor to measure inorganic phosphate sensorexploitingthe maltose phosphorylase, acid phosphatase, glucose oxidase and mutarotase to determine the concentration of phosphate. Researchers demonstrated that the sensor outputis well linear in the range 0.1 to1 μM, and sensor is now being used in environmental monitoring [45].

*2.7.3. Nitrate/nitrite.*Nitrate and nitrite are the important environmental contaminants causing environmental pollution. Various environmental agencies including US-EPA and EU authorities promulgated standards for drinking water. Biosensors found to be promising tool to detect and monitor level of nitrate and nitrite. A biosensor was developed by exploiting immobilized denitrifying bacteria for the determination of NO^{-3} in water samples. In a reaction chamber bacterial mediated reduction of NO^{-3} form N_2O occurs and it is measured by aspecific nitrous oxide micro-electrode, which is the main sensing part of the sensor [14]. New improvement in nitrate biosensor utilizes a whole-cell fluorescence system which is based on recombinant E Coli bacteria without the interference of phosphate, chloride and nitrite [18]. Nitrate biosensor based on electro-chemical technique also developed and provides more reliable and rapid detection of nitrate [46].

3. Conclusions

Application of biosensors is enormously increasing in many areas of environmental monitoring and assessment. However,some limitations such as selectivity, sensitivity, response time and usable lifetime is of great concern and vital study area for researchers. There is a need to improve the present biosensors and make themdesirable analytical tool in all aspects. The most probable areas of improvements that are expected to have a significant change and enhance the application of biosensor technology are: immobilization systems, nanotechnology, compactness, and multisensory array fortitudes for multiple contaminants determination [45]. Furthermore, there is a need to develop advanced biosensors based on modified base component and sensing elements which should be reliable, simple inmanufacturing and having the competence to widen the spectra of selectivity.

Analysis of complex real samples is possible with the development of advanced receptors which will support the in situ measurements eliminating the responses from nonspecific background effects. Present study revealed that the future of biosensors is promising however it depends on the proper implementation of evolvingerudite micro and nanotechnologies, chemistry, biochemistry, doping physics, and applied electronics. Biosensors development is a continuous process and there is always room for improvement in the design and components of emerging biosensors. With the addition of more environmental contaminants which may be threat for humans and the overall ecosystem, the need for faster and more accurate biosensors will always at high-rise. Biosensors at presentrequired to get the trust of potential consumers, keeping in mind that the acceptance of new manoeuvresis the best indicator of the success and achievement for an emerging technology [14].

References

[1] Piliarik, M., Šípová, H., Kvasnička, P., Galler, N., Krenn, J. R., & Homola, J. (2012). High-resolution biosensor based on localized surface plasmons. *Optics Express*, *20*(1), 672-680.

[2] Xiao, Y., Lubin, A. A., Heeger, A. J., & Plaxco, K. W. (2005). Label-free electronic detection of thrombin in blood serum by using an aptamer-based sensor. *Angewandte Chemie*, *117*(34), 5592-5595.

[3] Krishnamurthy V, Monfared S, Cornell B (2010). ": Ion Channel Biosensors Part I Construction Operation and Clinical Studies". *IEEE Transactions on Nanotechnology* 9 (3): 313–322.

[4] Anthony P. F. Turner (2013). Biosensors: sense and sensibility *Chem. Soc. Rev.*, 42(1): 3184-3196.

[5] Palecek, E., and F. Jelen. (2002). Electrochemistry of nucleic acids and development of DNA sensors. Crit. Rev. Anal. Chem. 3:261-270.

[6] Phoenix, P., Keane, A., Patel, A., Bergeron, H., Ghoshal, S., & Lau, P. C. K. (2003).

Characterization of a new solvent-responsive gene locus in Pseudomonas putida F1 and its functionalization as a versatile biosensor. *Environmental microbiology*, *5*(12), 1309-1327.

[7] Drummond, T. G., Hill, M. G., & Barton, J. K. (2003). Electrochemical DNA sensors. *Nature biotechnology*, *21*(10), 1192-1199.

[8] Albers, J., T. Grunwald, E. Nebling, G. Piechotta, and R. Hintsche.2003. Electrical biochip technology—a tool for microarrays and continuous monitoring. Anal. Bioanal. Chem.377:521-527.

[9] Albers, J., Grunwald, T., Nebling, E., Piechotta, G., & Hintsche, R. (2003). Electrical biochip technology—a tool for microarrays and continuous monitoring. *Analytical and bioanalytical chemistry*, *377*(3), 521-527

[10] Hung, Y. P., Albeck, J. G., Tantama, M., & Yellen, G. (2011). Imaging Cytosolic NADH-NAD< sup>+</sup> Redox State with a Genetically Encoded Fluorescent Biosensor. *Cell metabolism*, *14*(4), 545-554.

[11] Cavalcanti, A., Shirinzadeh, B., Zhang, M., & Kretly, L. C. (2008). Nanorobot hardware architecture for medical defense. *Sensors*, *8*(5), 2932-2958

[12] Turner, A. P. F., & Piletsky, S. (2005). Biosensors and Biomimetic Sensors for the Detection of Drugs, Toxins and Biological Agents. In *Defense against Bioterror* (pp. 261-272). Springer Netherlands.

[13] Gooding, J. J. (2002). Electrochemical DNA hybridization biosensors. *Electroanalysis*, *14*(17), 1149-1156.

[14] Rodriguez-Mozaz, S., De Alda, M. L., Barceló, D., & Marco, M. P. (2004). Biosensors for environmental applications: future development trends. *Pure and Applied Chemistry*, *76*(4), 723-752.

[15] Urban, G. A. (2009). Micro-and nanobiosensors—state of the art and trends. *Measurement Science and Technology*, *20*(1), 012001.

[16] Scognamiglio, V., Pezzotti, G., Pezzotti, I., Cano, J., Buonasera, K., Giannini, D., & Giardi, M. T. (2010). Biosensors for effective environmental and agrifood protection and commercialization: from research to market. *Microchimica Acta*, *170*(3-4), 215-225.

[17] Wang, J., Rivas, G., Cai, X., Palecek, E., Nielsen, P., Shiraishi, H., ... & Flair, M. N. (1997). DNA electrochemical biosensors for environmental monitoring. A review. *Analytica Chimica Acta*, *347*(1), 1-8.

[18] Rodriguez-Mozaz, S., de Alda, M. J. L., & Barceló, D. (2006). Biosensors as useful tools for environmental analysis and monitoring. *Analytical and bioanalytical chemistry*, *386*(4), 1025-1041.

[19] Kostrzynska, M., Leung, K. T., Lee, H., & Trevors, J. T. (2002). Green fluorescent protein-based biosensor for detecting SOS-inducing activity of genotoxic compounds. *Journal of microbiological methods*, *48*(1), 43-51.

[20] Lee, J. H., & Gu, M. B. (2005). An integrated mini biosensor system for continuous water toxicity monitoring. *Biosensors and Bioelectronics*, *20*(9), 1744-1749.

[21] Ouyang, H., Striemer, C. C., & Fauchet, P. M. (2006). Quantitative analysis of the sensitivity of porous silicon optical biosensors. *Applied Physics Letters*, *88*(16), 163108-163108.

[22] Tombelli, S., Minunni, M., & Mascini, M. (2005). Piezoelectric biosensors: Strategies for coupling nucleic acids to piezoelectric devices. *Methods*, *37*(1), 48-56.

[23] Zhang, Y., & Tadigadapa, S. (2004). Calorimetric biosensors with integrated microfluidic channels. *Biosensors and Bioelectronics*, *19*(12), 1733-1743.

[24] Brack, W., Ulrich, N., & Bataineh, M. (2011). Separation Techniques in Effect-Directed Analysis. In *Effect-Directed Analysis of Complex Environmental Contamination* (pp. 83-118). Springer Berlin Heidelberg.

[25] Yamaguchi, A., Ishibashi, H., Kohra, S., Arizono, K., & Tominaga, N. (2005). Short-term effects of endocrine-disrupting chemicals on the expression of estrogen-responsive genes in male medaka (< i> Oryzias latipes</i>). *Aquatic toxicology*, *72*(3), 239-249.

[26] Philp, J. C., Balmand, S., Hajto, E., Bailey, M. J., Wiles, S., Whiteley, A. S., ... & Dunbar, S. A. (2003). Whole cell immobilised biosensors for toxicity assessment of a wastewater treatment plant treating phenolics-containing waste. *Analytica Chimica Acta*, *487*(1), 61-74.

[27] McKnight, T. E., Melechko, A. V., Griffin, G. D., Guillorn, M. A., Merkulov, V. I., Serna, F., ... & Simpson, M. L. (2003). Intracellular integration of synthetic nanostructures with viable cells for controlled biochemical manipulation. *Nanotechnology*, *14*(5), 551.

[28] Ptitsyn, L. R., Horneck, G., Komova, O., Kozubek, S., Krasavin, E. A., Bonev, M., & Rettberg, P. (1997). A biosensor for environmental genotoxin screening based on an SOS lux assay in recombinant Escherichia coli cells. *Applied and environmental microbiology*, *63*(11), 4377-4384.

[29] Rodriguez-Mozaz, S., Marco, M. P., de Alda, M. J. L., & Barceló, D. (2004). Biosensors for environmental monitoring of endocrine disruptors: a review article. *Analytical and bioanalytical chemistry*, *378*(3), 588-598.

[30] Rier, S., & Foster, W. G. (2003). Environmental dioxins and endometriosis. In Seminars in reproductive medicine (Vol. 21, No. 02, pp. 145-154). Copyright© 2003 by Thieme Medical Publishers, Inc., 333 Seventh Avenue, New York, NY 10001, USA. Tel.:+ 1 (212) 584-4662.

[31] Götz, R., Sokollek, V., & Weber, R. (2012). The dioxin/POPs legacy of pesticide production in Hamburg: part 2—waste deposits and remediation of Georgswerder landfill. Environmental Science and Pollution Research, 1-12.

[32] Chen, H., & Zhuang, H. (2012). Determination of 3, 4-dichlorinated biphenyl in soil samples by real-time immuno-PCR assay. *Journal of Environmental Sciences*, *24*(12), 2191-2197.

[33] Kasai, A., Hiramatsu, N., Meng, Y., Yao, J., Maeda, S., & Kitamura, M. (2005). Fast-track DRESSA: a bioassay for fast, sensitive, and selective detection of halogenated and polycyclic aromatic hydrocarbons. *Analytical biochemistry*, *337*(1), 84-88

[34] Parellada, J., Narvaez, A., Lopez, M. A., Domınguez, E., Fernandez, J. J., Pavlov, V., & Katakis, I. (1998). Amperometric immunosensors and enzyme electrodes for environmental applications. *Analytica chimica acta*, *362*(1), 47-57.

[35] Filik, H., Hayvalı, M., Kılıç, E., Apak, R., Aksu, D., Yanaz, Z., & Çengel, T. (2008). Development of an optical fibre reflectance sensor for< i> p</i>-aminophenol detection based on immobilised bis-8-hydroxyquinoline. Talanta, 77(1), 103-109..

[36] Servos, M. R., Bennie, D. T., Burnison, B. K., Jurkovic, A., McInnis, R., Neheli, T., ... & Ternes, T. A. (2005). Distribution of estrogens, 17β-estradiol and estrone, in Canadian municipal wastewater treatment plants. *Science of the total environment*, *336*(1), 155-170.

[37] Rodriguez-Mozaz, S., Marco, M-P., de Alda, M. J. and Barceló, D. (2005). A global perspective: Biosensors for environmental monitoring, *Talanta*. 65(1), 291-297.

[38] Rodriguez-Mozaz, S., Marco, M. P., de Alda, M. J., & Barceló, D. (2004). Biosensors for environmental applications: Future development trends. *Pure and applied chemistry*, *76*(4), 723-752.

[39] Hansen, L. H., & Sørensen, S. J. (2000). Detection and quantification of tetracyclines by whole cell biosensors. *FEMS microbiology letters*, *190*(2), 273-278.

[40] Arora, P., Sindhu, A., Kaur, H., Dilbaghi, N., & Chaudhury, A. (2013). An overview of transducers as platform for the rapid detection of foodborne pathogens. *Applied microbiology and biotechnology*, *97*(5), 1829-1840.

[41] Ibarra-Escutia, P., Gómez, J. J., Calas-Blanchard, C., Marty, J. L., & Ramírez-Silva, M. T. (2010). Amperometric biosensor based on a high resolution photopolymer deposited onto a screen-printed electrode for phenolic compounds monitoring in tea infusions. *Talanta*, *81*(4), 1636-1642.

[42] Chouteau, C., Dzyadevych, S., Durrieu, C., & Chovelon, J. M. (2005). A bi-enzymatic whole cell conductometric biosensor for heavy metal ions and pesticides detection in water samples. *Biosensors and Bioelectronics*, *21*(2), 273-281.

[43] Ongaro, M., & Ugo, P. (2012). Bioelectroanalysis with nanoelectrode ensembles and arrays.

Analytical and bioanalytical chemistry, 1-15.

[44] Parellada, J., Narvaez, A., Lopez, M. A., Domınguez, E., Fernandez, J. J., Pavlov, V., & Katakis, I. (1998). Amperometric immunosensors and enzyme electrodes for environmental applications. *Analytica chimica acta*, *362*(1), 47-57

[45] Conrath, N., Gründig, B., Hüwel, S., & Cammann, K. (1995). A novel enzyme sensor for the determination of inorganic phosphate. Analytica chimica acta, 309(1), 47-52.

[46] Cosnier, S., Da Silva, S., Shan, D., & Gorgy, K. (2008). Electrochemical nitrate biosensor based on poly (pyrrole–viologen) film–nitrate reductase–clay composite. Bioelectrochemistry, 74(1), 47-51.

IOP Conf. Series: Materials Science and Engineering **51** (2013) 012013 doi:10.1088/1757-899X/51/1/012013

Scope for biological sensing technologies in meat production and export in northern Pakistan

M S Qureshi[1] and I H Qureshi[2]
[1]Faculty of Animal Husbandry and Veterinary Sciences, The University of Agriculture, Peshawar-25120, Pakistan (E-mail: drmsqureshi@aup.edu.pk)
[2]Center for Advance Studies in Engineering, Islamabad, Pakistan.

Abstract.

The Khyber Pakhtunkhwa province of Pakistan is rich in livestock resources, including 14.84 million sheep and goats (valued at US$1.60 billion) and a 27% share of the national poultry sector (having an investment of US$2.00 billion), and produces 834 billion kg meat. These huge assets have the potential to support the provincial economy through income generation, self employment and production of certified high-quality food items for the domestic and international Halal Food Market. A model has been developed for analyzing the gaps in the status of health, productivity, nutrition, fertility and management aspects of local farming. Improved practices would be introduced to combat the losses. The model will comprise a farming network linked to farmers' welfare centre, a central lab and an expert group. A strong sensing technology network would be introduced for data transfer and quality control of the inputs and products. The farmers will e-tag their animals for the purpose of traceability, online history and biodata. The data will be maintained in remote and central servers. A communication system would be developed utilizing mobile phones for the prices, demands and availability status of inputs and produce at local and international markets. A mobile money transfer system will be introduced to exchange, save and borrow small amounts of capital as well as take out short-term insurance policies.

Introduction

Livestock contributed approximately 55.4 percent to the agricultural value added and 11.9 percent to national GDP during 2012-13, against 55.3 percent and 11.9 percent during the same period last year. Gross value addition of the livestock sector at constant cost factor has increased up to Rs.756 billion during 2012-13. Cattle, buffaloes, sheep and goats are major food animals producing 49.5 billion kg milk and 3.4 billion kg meat, with a value of US$ 28.5 billion per annum. The population of the four species of animals showed respective growth of 141, 143, 61 and 269%, with goats recording the highest growth rate, reaching the present population of 58 million heads [1] (Figure 1). A significant share of meat produced in the country is contributed by sheep and goats (19%, Figure 2).

[1] To whom any correspondence should be addressed.

THE WORLD MEAT TRADE

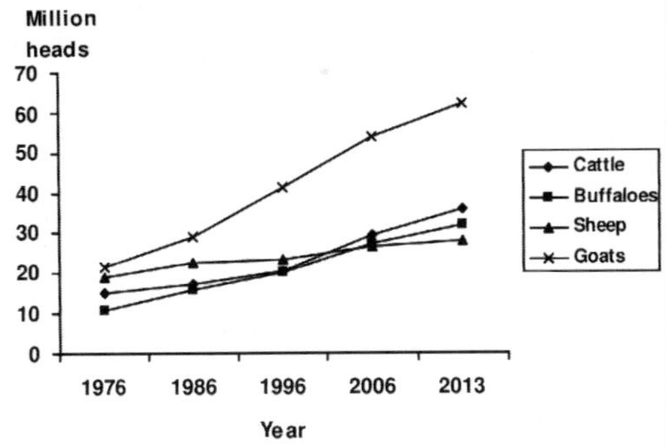

Figure 1. Change in livestock population

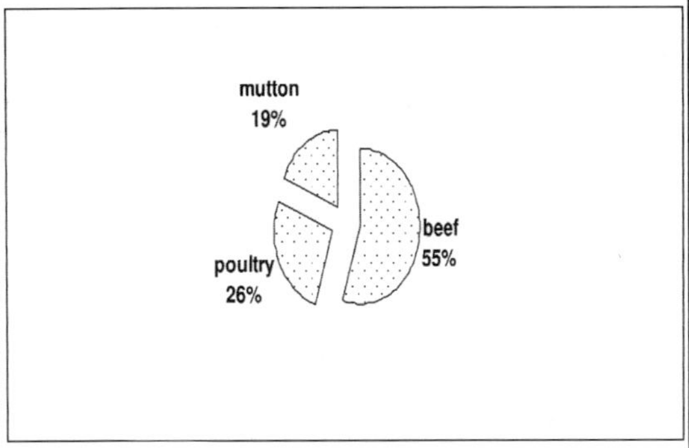

Figure 2. Meat production during 2012-13 (3.23 billion kg)

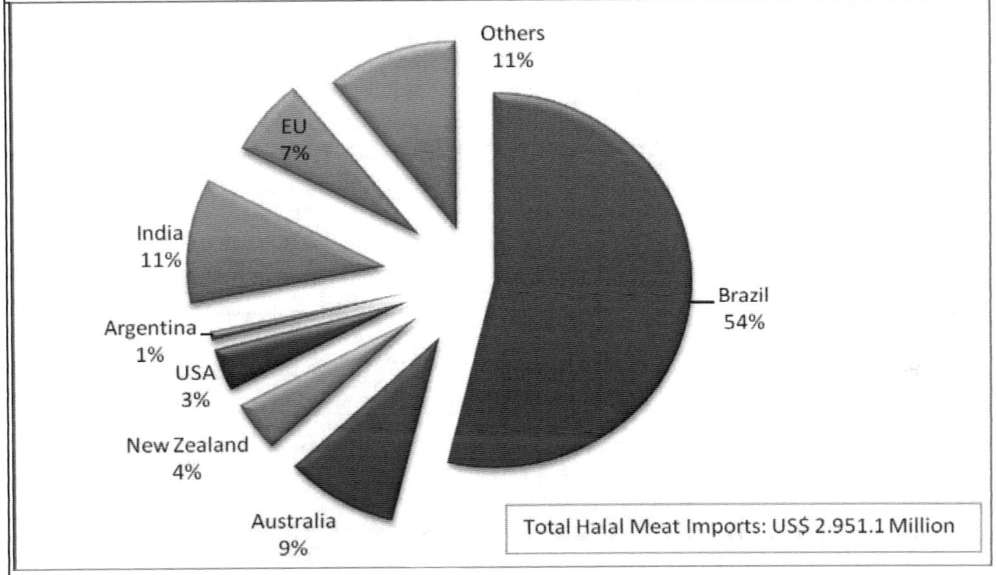

Figure 3. Import of Halal meat to the Gulf Cooperation Council Region

The export of Halal meat to the Gulf Cooperation Countries has been reported as US$ 2.95 billion. This meat has been produced in the non Muslim regions of Brazil, India, Australia, EU, New Zealand and the USA [2] (Figure 3). Halal International Halal Integrity (IHI) Alliance – Global Halal Authority has been constituted as part of World Halal Forum resolution 2007 and OIC General Assembly 2008. Malaysia approved 45 Halal certification Bodies, Indonesia 40 and GCC countries 52. IHI has been created to serve as a platform for providing services to its members, representing their interests and to liaise with governments, organizations and business communities, including in advising on Halal certification & related matters (www.ihialliance.org). A partnership agreement was signed with Islamic Chamber of Commerce & Industry under the leadership of it's President, Sheikh Saleh Abdullah Kamel from Saudi Arabia on 12 May 2008.

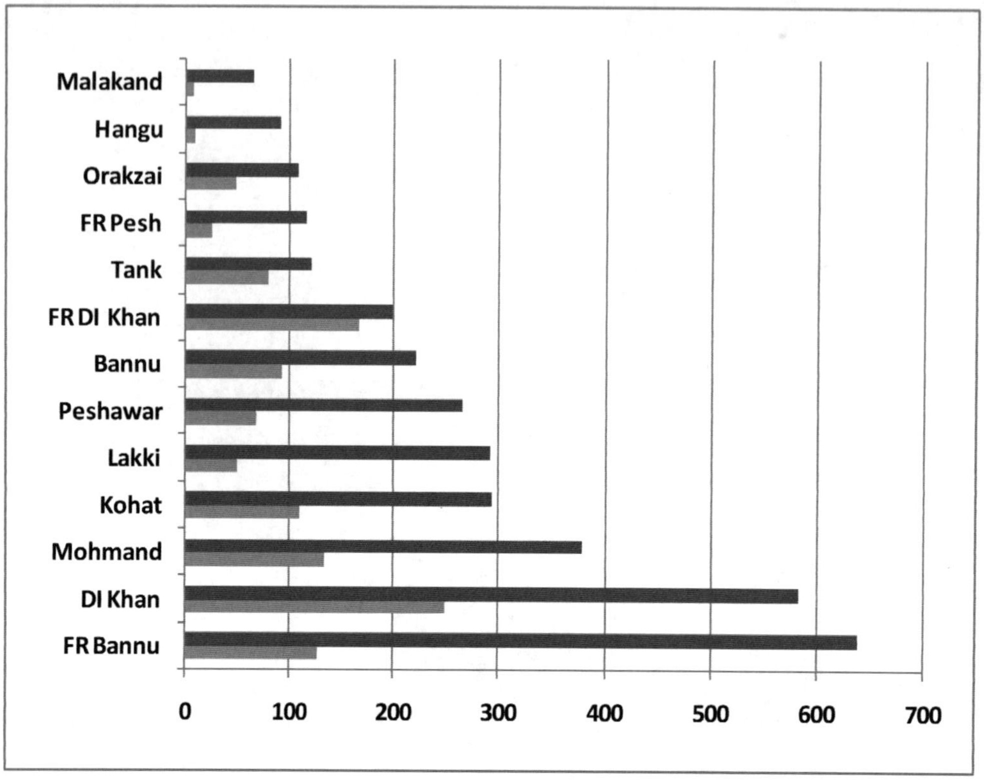

Figure 4. Sheep and goats population in the northern Pakistan (Goats population in million heads, is given in red and Sheep in green; extracted from Livestock Census, 2006)

The Khyber Pakhtunkhwa (KP), the Northern Province of Pakistan with its arid and hilly nature inhabits majority of the small ruminants. These animals are kept by the farmers on small scale under extensive farming system. Sheep and goats, being small-sized ruminants, are capable of integrating into the dissimilar socio-economic situations that is prevailing in Pakistan. Population of sheep and goats in various districts and tribal areas of the province is given in Figure 4. The province has got a great potential for meat export. The federal government of Pakistan is presently executing several projects in livestock sector at national level with an estimated cost of Rs 8.8 billion [1]. These projects focus on promoting milk and meat production/marketing; strengthening of extension services and delivery system to livestock farmers; prevention and control of livestock and poultry diseases; up-gradation of animal quarantine services and provision of veterinary services at farmer's door step. During the 2009 to 2010 period, technical and financial assistance were provided to farmers, totaling 13,171 fattening operations involving 381 678 animals under the Meat Development Project. Goat has been an important species of these development programs.

MEAT PRODUCTION AND EXPORT POTENTIAL

The total meat produced in the country has been 2.23 billion kg per annum as per latest economic survey [1], comprising 55, 26 and 19% beef, poultry and mutton respectively. This production level has been achieved in spite of the fact that the animals are not provided

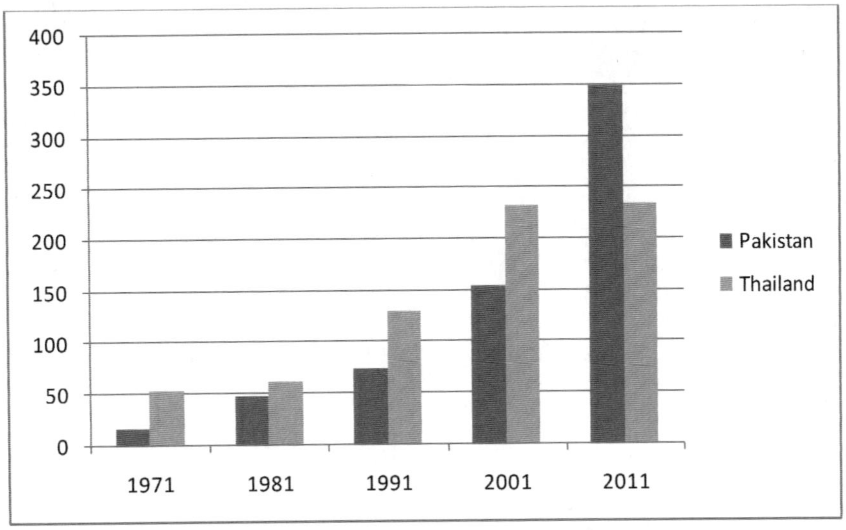

Figure 5. Change in poultry population (million heads) in Pakistan and Thailand; Thailand's poultry meat export reached 1.75 billion US$ verses nil for Pakistan, FAOSTAT 2013)

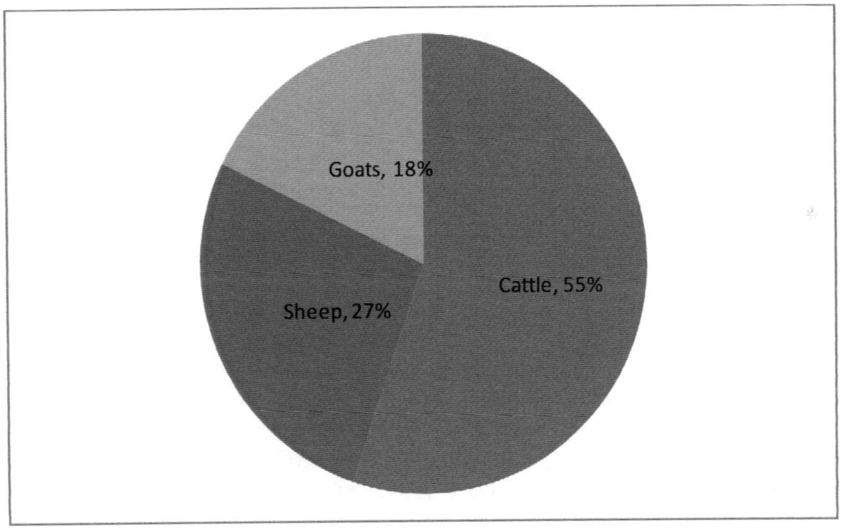

Figure 6. Export of meat of various animals from Pakistan during 2010 (total value US$ 93.08 million, FAOSTAT 2013); The huge meat produced by poultry industry was 0.83 billion kg which was ignored.

sufficient feeding, health and marketing inputs, impeding their productivity and profitability. However, the huge potential of the northern areas of Pakistan has got the potential to play their due role in generating self employment for the war hit region and surplus food for earning foreign exchange.

Pakistan has got several goat breeds [3]. Beetal goat is the most popular milk breed, found in the central Punjab, possessing a massive head, Roman nose, long, broad and pendulous ears, well-developed udder and long teats. Milk yield of Beetal goats has been recorded at 190 L in 150-day lactation period. The goat is fertile with more than 50% twining or triplet births. Male Beetal goats having body weights of 70 to 80 kg are being raised especially for sacrifice on Eid-ul-Azha. Dera Din Panah goat is found in Muzaffargarh and Multan Districts. It has a large head and Roman nose. Milk yield has been recorded at 160 L in 150-day lactation period. Twin births are common in this breed. Hairy goat is a milk breed while Kajli (Pahari) goat is a meat breed of Southern Punjab. Nachi and Pothowari goats are meat breeds of Southern and Teddy is of the Northern Punjab. In the Sindh province, Chappar, Bari, Bugri, Tapri and Desi meat breeds and Jattan, Kamori and Pateri milk breeds are found.

Damani is a dual purpose while Gaddi and Kaghani are meat breeds of the Khyber Pakhtunkhwa province. In Baluchistan, Kurasani goat is a dual purpose breed and Lehri is a meat breed. Goats survive well under the rural environment. They are acclimatized to the diverse agro-climatic conditions and manifest higher fertility and short generation interval, and thus are the animal of farmers' choice. Because of their low maintenance cost, quick return on capital and low capital investment risk, goats are ideally suited for the poor rural folk especially the marginal and landless laborers. To cut cost of production, goats are usually taken care by engaging family members, especially children and women.

Figure 5 reflect change in poultry population (million heads) in Pakistan and Thailand; Thailand's poultry meat export reached 1.75 billion US$ verses nil for Pakistan, FAOSTAT 2013). Pakistan performed parallel to Thailand in increasing poultry production; and even exceeded during the last decade; however it could not proceed parallel in respect of making preparations for exporting the produce. As evident from Figure 6 the export of meat of various animals from Pakistan during 2010 [4] (total value US$ 93.08 million) was contributed by cattle, sheep and goats. The huge meat produced by poultry industry was 0.83 billion kg was ignored. The export was materialized by Punjab Meat Company and some private firms.

DAIRY SCIENCE PARK

The International Workshop on Dairy Science Park was held at the Agricultural University Peshawar [5]. The conference was attended by more than 450 delegates from all the four provinces of the country belonging to a variety of segments of the society. Various activities have been proposed for productivity enhancement and industrial applications. Some projects have been identified for implementation at Agricultural University Peshawar. Besides the support already provided to the Faculty of Animal Husbandry and Veterinary Sciences, this University has started dialogues for initiating some additional activities. A liaison has been established with the Khyber Pakhtunkhwa Chamber of Commerce and Industry for supporting commercially viable projects.

The provincial government and other local and international donors are expected to sponsor some viable projects. Goats are being considered as small enterprises targeted at self employment, food security and export to the Halal food market. Local and international investors are being invited to establish partnership with the Dairy Science Park. All the components of the Park would be integrated through a network getting due inputs from the farmers, students, veterinarians, investors, engineers, service providers, inputs suppliers and exporters.

POTENTIAL ROLE OF SENSING TECHNOLOGY

Geographic Information System

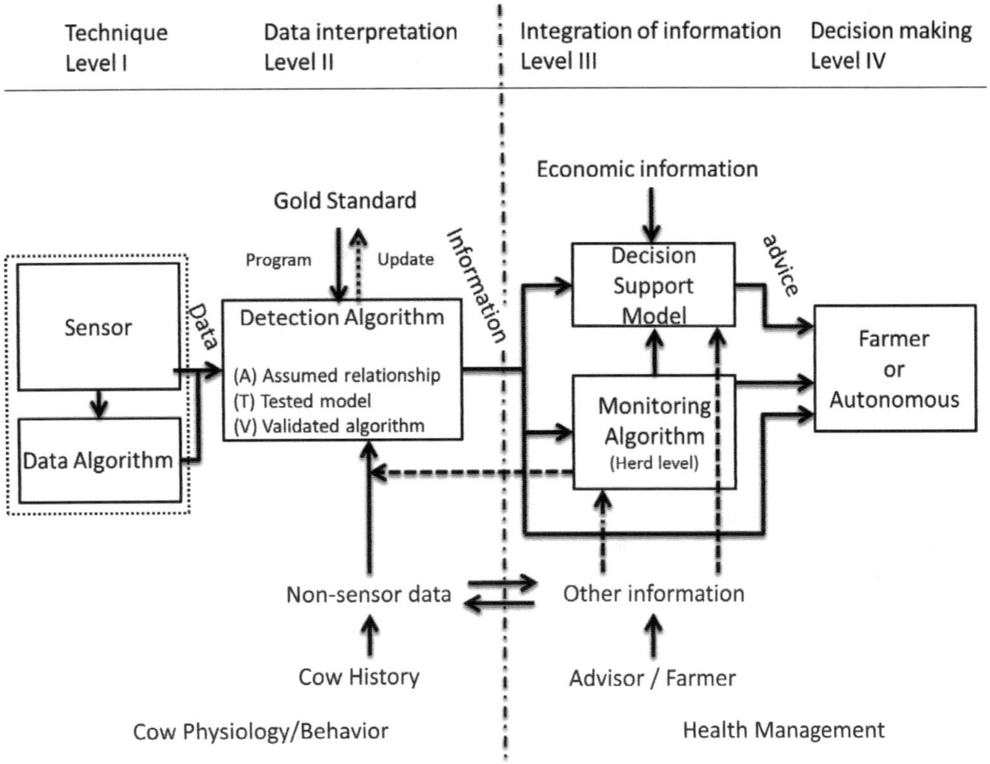

Figure 7. Framework of the use of sensor information in dairy farm management.

Geographic information system (GIS) will be introduced in the region which is a system designed to capture, store, manipulate, analyze, manage, and present all types of geographical data [6]. It may help in assessing the physical and biological coverage of land and generate information on a combinations of: i) vegetation types/pastures/rotational grazing; ii) soils/degradation/erosion; iii) water bodies; iv: agriculture/at risk of conversion/deforestation and; v: built environments. The shepherds may be provided information on availability and carrying capacity of green masses for grazing of their animals. In the meantime information may be provided to the District Livestock Officers for arranging vaccinations, deforming and making quarantine and marketing armaments for the incoming and outgoing animals from their respective regions. The pasture areas are consistently decreasing in various regions of KP as the water table is going down with the passage of time. This results in decrease in vegetation which leads to further degradation of land. So a vicious circle is constituted where the decrease in vegetation and land degradation support and enhance each other.

E-tagging and animal tracking

Various sensing systems have been introduced like GG02 Mini Solar GPS Tracker, which helps in identification and tracking of the animals. E-tagging has been introduced at livestock farms for the purpose of: i) traceability; ii) online history and; iii) biodata, data transfer for management intervention. The data is maintained in remote and central servers. A communication has been developed utilizing mobile phones for the prices, demands and availability status at local and international markets for inputs and products.

Environment monitoring

Altimeter/barometer has been used for assessing animal's physiological requirements comprising assessing stress and economic productivity. Environment meter has been developed for measuring comfortably of animals, ambient Temperature, humidity, light intensity and wind speed. The device reported has been m-9100 Lutron Taiwan. Portable CO2 meter has been developed like Model : GCO-2008 for measuring CO (Carbon monooxide) and temperature with a response time of 30 seconds.

Barometer Weather Station have been in use for weather forecasting being sunny or slightly cloudy. Digital barometer 800 mbar to 1050 mbar with an air pressure (mb/hpa or inhg unit select), operating altitude 300 ~ +1200M; 5), past 12 hour pressure history bar graph and measuring temperature range : -9¡æ to +70¡æ

Application of mobile phones have been used due to timeliness of data, accessibility/availability and accuracy of data. The data may be extracted from cell location/GPS, photo/video/sound or bar code reader. The data is user based, is cost effective and is superior to computer due to its portability, easiness an networking. A mobile money transfer systems has been developed for: i) exchange; ii) saving; iii) borrowing small amounts of capital and; iv) short-term insurance policie [7].

Health sensors

Since the 1980s, work has been done on devices that measure a health indicator in, up, on, or from an individual cow [8]. Examples of sensors include milk electrical conductivity, milk color sensors, acceleration sensors (attached to the cows leg and pH sensors. A sensor system consists of the device itself plus the software that processes the data to produce information or advice. Figure 7 has reproduced a framework of the use of sensor information in dairy farm management referred to above. The scheme consists of four levels of activities comprising selection of the activity to be recorded, data interpretation, integration of information and decision making.

References

[1] Economic Survey, 2013. Government of Pakistan, Economic Affairs Division, Islamabad.

[2] The World Bank First EAP Regional Agribusiness Trade and Investment Conference "Agro-enterprise Without Borders". Singapore, 30 – 31 July 2009

[3] Khan, B.B., A. Iqbal and M.I. Mustafa, 2003. Sheep and goat. Production. Department of Livestock Management, University of Agriculture Faisalabad

[4] FAOSTAT 2013. Food and Agriculture Organization of the UN, Rome

[5] Qureshi MS, Proc *Int Workshop Dairy Sci Park* - Peshawar - Nov 21-23. Managing editor. Agric Univ Peshawar, Pakistan, 2011

[6] Marsett RC, J Qi, P Heilman, SH Biedenbender, M CarolynWatson, S Amer, M Weltz, D Goodrich and R Marsett 2006. *Rangeland Ecol Manag* 59, 530-40

[7] FAO news article. Cell phones revolutionizing Kenya's livestock sector, 1 March, 2013, fao.org

[8] Rutten C J, Velthuis A G J, Steeneveld W, and Hogeveen H *J Dairy Sci.* **96** :1928–52

Biosensors for termite control

M Farkhanda[1]
Department of Zoology, Lahore College for Women University, Lahore, Pakistan

E-mail:doc_farkhanda@yahoo.com

Abstract: Termites are major urban pests in Pakistan and cause damage to wooden structures and buildings. Termite management has two parts: prevention and control. The most difficult part of termite control is termite detection as most of them are subterranean in Pakistan and have tunneling habit.Throughout the world, chemical termiticides are going to be replaced by baits, microwave and sensor technology. Termite species are distinct biologically and have specific foraging behaviors. Termite Detection Radar, Moisture meter and Remote Thermal Sensor with Laser are available throughout the world. These can detect termites underground and use fewer chemicals than traditional methods. For wooden buildings, a termite sensor and an intrusion detection system for detecting termites are designed. A pair of electrodes is disposed inside the container. A pair of terminals is connected to these electrodes, these extend outside the container. Termites are detected by a change of conductivity between the electrodes, when termites are detected a warning device generates a warning signal. In Pakistan, there is dire need to develop such biosensoring devices locally, then apply control methods that would save money and protect the environment.

1. Introduction

Termites are the most important group of insects that destroy cellulosic materials including wood. These social insects having soldier, worker, queen and king caste,are widespread to terrestrial environmentand are distributed in Tropical, Subtropical and Temperate regions(Fig 1). They cause destruction to power pole, railway sleepers, many crops plantation and living treesbut perhaps of greatest importance is the destruction inflicted on timber used in constructional purposes both outdoors and inside buildings. In Pakistan, there are 53 species of termites [1]. During survey thirteen termite species have been identifiedas cost-effective species to the residential wooden structures in Pakistan. Some of these are *Heterotermesindicola, Coptotermesheimi, Microtermesmycophagus, Microtermesobesi, Eremotermesparadoxalis, Odontotermeshorai, Odontotermesgurdasurensis, Odontotermesguptai, Odontotermesobesus*[2]. Worldwide the cost of treatments to eradicate subterranean termite infestations in 1986 has been estimated at approximately US$2 billion [3] but in Pakistan no such survey is available to determine the cost of damage.

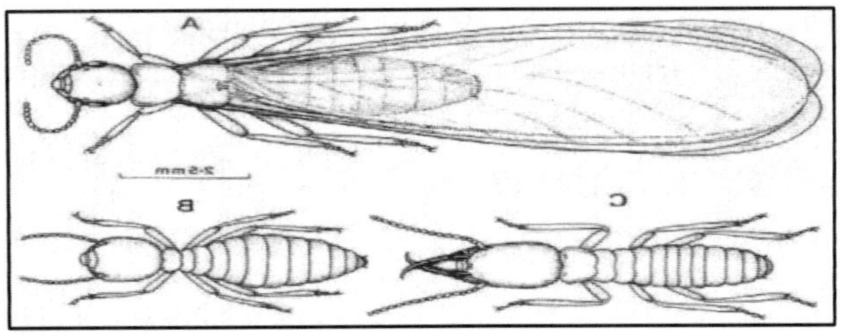

Fig1. Termite alate, worker and soldier. Source: CSIRO Division of Entomology 1991

The most common traditional termite control techniques are soil chemical barrierspre-construction and post construction treatment. Besides this termite baiting systems, Physical treatment to aggregate barriers such as stainless steel mesh barriers and Plastic barriers are also used to treat termites in different parts of the world.

2. Results and Discussion

Now the question is why we should use biosensors? The answer is very simple. The biosensors do no help to control termites but help only to detect termites. It gives clarity about activity of termite attack and decrease tension between parties and prevent unnecessary treatments. One more reason is biosensors are used for termite detection in wood due to following wood properties: wood is differentiated as hardwood and softwood, It has dielectric and conductivity properties of wood and show microwave effect. As termites show foraging behavior and go on moving in search of food, this technology also defines the frequency range and wave length of the signals that are generated by termite noises [4].

Currently, three remarkable techniques to identify trace and confirm thetermitesinfestation without substantially searching and destruction of building structures, these are:

- Thermal Imaging
- Moisture Meters
- Termite Detection Radar

2.1.Thermal Imaging

Thermal imaging can be used as a non-destructive and fast method to detect termites in trees and buildings compared to the traditional methods such as knocking and drilling in wood. During thermal imaging,thermal camera creates images based on thermal energy instead oflight, the subsequent image appears asan X ray containing heat regions. As we know termite activity in nest generates heat that is detected by infra-red camera, so it helps in termite detection in behind walls and other surfaces. This two-dimensional temperature mapping technique has potential for characterizing products during several operations of agricultural and food industries. Thermal imaging has been successfully adopted for studying plant physiology, irrigation scheduling, and yield forecasting in agricultural fields.

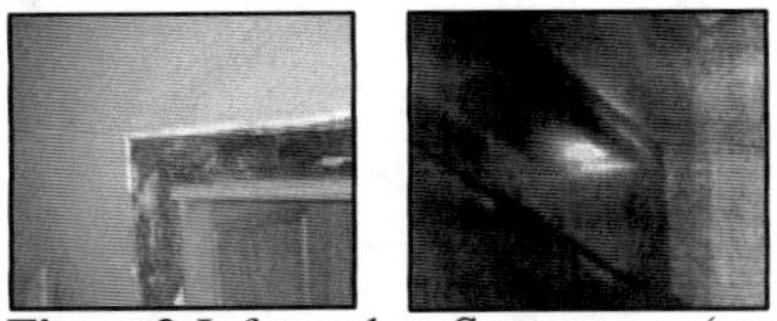

Figure2: Infra-red Systems (e.g. Thermal imaging) Source: James and Rice 2002

*2.1.1. Thermal Sensor with Laser Guide.*Thermal sensor containing laser guide senses possible termite infested locations through warmer areas. Heat and humidity produced by the subterranean termites in the adjacent environment are the two major factors supporting the operator to detect potential termite threat points by indicating heat differentialzones. As there are temperature readings on equipment, so it allows operator to determine high risk areas where termite is present. Similarly, during Acoustic emission, there is rapid localized release of energy and it generates an elastic wave which travels through the material and converts the (ultra)sound waves into small variations of voltage, which can be measured. Drywood termites are normally detected with a small handy device throughtheir entrance in structures, mud tubes and nests without causing any physical damage to structures [5].

2.2.Moisture Meter
Another biosensor device is Moisture Meter which pinpoints the concentrated high moisture level areasto identify termite risk locations. This device identifies,measures and records different moisture levels in building materials without penetrating the buildings physically.

2.3.Termite Detection Radar
These radars are used for accurate detection and confirmation of termite presence.

Resistivity imaging survey, Tomography scanning and Millimeter-wave imaging techniques are the modern techniques to detect and control termites. By using these biosensor techniques, it would be far more economical to protect the timber-in-service by using Termite Management Systems (TMS) by sensing a combination of factors such as temperature, moisture, movement, feeding and behaviour of the termites [5] [6].

Another useful tool is boroscope, that would drill a small hole in the wall and the scientists can have a look on termites inside walls and then decide treatment. Besides termites there are also Decay, Fungi and wood borers that are detected by biosensors.

3.Conclusions
In future, we should focus on research that is to examine, screen, develop and refine a non-destructive electronic / microwave techniques for the detection and location of termite damages in wood and at buildings. These biosensor devices would be a combination of factors such as temperature, moisture, movement, feeding and behaviour of the termites in sensing technology.

4.References
[1] Manzoor F 2006 *Morphometric studies on the termite genus* Odontotermes. Monograph published byH.E.C.Islamabad.
[2] Manzoor F and Mir N 2010Survey of Termite Infested Houses, Indigenous Building Materials and Construction Techniques in Pakistan *Pak. J. Zool.* **42***(6)* **693-696**.
[3] Edwards R and MillA E 1986: *Termites In Buildings- Their Biology & Control.* Rentokil Ltd.
[4] (Pre-Amplifier, Sound Amplifier and Sound Level Meter) Source:Electronics 1994) Practical Experiments
[5] Acoustic Emission Technology (e.g. PICUS® Sonic Tomograph) Source: PICUS 2007
[6] Source: Termite Detection Systems 2001
[7] Source: Standards Australia Group 1999
[8] http://www.termatrac.com/learn-about13/termite-detection-radar-technology17
[9]http://conquertermites.com.au/index.php/uncategorized/termite-inspections-meet-some-of-the-tools-of-

A Simple Method for the Calibration of an Open Surface Water Bath

Muhammad Ghazanfar Ali
Indus University, Karachi
PCSIR Lab Complex Karachi
E-mail: mghazanali@yahoo.com

Abstract. Calibration and testing of open surface water bath is a new technique and may be used for other liquid bath testing. Due to some critical circumstances the bath is not covered by the manufacturing company so is prone to the effects of surrounding environmental conditions. Due to this effect, bath temperature is not stable and calibration and testing cannot be performed under these conditions. For this type of calibration the reference instruments should measure temperature with very high levels of accuracy. A Platinum Resistance Thermometer (PRT), with super thermometer (indicator), is used for high accuracy calibration. Oil-filled flasks are used as a secondary medium to calibrate the bath and to reduce the evaporation, and instability in the temperature measurements. Therefore, on the basis of stable temperature with good precision and minimum uncertainty, the proposed method is found to be suitable for the calibration of open surface water baths.

Keywords: *Calibration, water bath, uncertainty measurement*

1. Introduction

Open surface water baths are commonly used in a number of applications in all fields of science and technology especially in medical and textile laboratories. Temperature, humidity, atmospheric pressure, and dust particles are the main factors which adversely affect the accuracy of water bath's temperature. Therefore, water bath needs calibration prior to use. Unfortunately, a little data is available in which calibration of such type of equipment is considered.

Calibration is a process in which, the performance of an instrument is compared against standards of known uncertainty and error to detect or correct / adjust any variation from the required performance specifications. In other words, it is a process of verifying the operational integrity of an instrument.

In this study three different types of oil were used as a secondary medium to calibrate the bath to reduce the evaporation, and instability in the temperature measurements of water. Traceability in the measurements is achieved using a standard reference thermometer, whose value is traceable to the NIST USA. Generally, the fitness of the proposed or newly developed methods is assessed through method validation studies. As there is a lack of a simplified standard method for the calibration of such type of bath; the compliance of the proposed method was assessed by uncertainty measurement. The data was produced on overall performance of water bath. The individual factors which influenced the uncertainty associated with the results of the proposed method were identified and corrected.

The aim of the present work is to propose a novel, simple and accurate method for the calibration of open surface water baths, which is special type of water bath use in chemical, Medical and textile industries where Thousands of important tests are performed with the help of water bath. So the proposed method is a technique to check and verify the performance of open surface water bath which is the basic requirement of ISO (International organization for standardization) clauses.

2. Experimental Work

2.1. Materials

PRT (Platinum Resistance Thermometer) with super thermometer (indicator) cropico made was used as a secondary standard reference device. PRT has secondary Standard sensor according to ITS-90. And supper thermometer indicator is highly precise device having three decimals readability, which is more accurate as compare to bath. Three types of oil (transformer oil, cooking oil, Hello carbon 0.8 grades) were selected as a reference medium for temperature measurements. All samples of oil were of commercial grade and purchased from local markets.

Selection of oil is very important because flashing point is considered. Flash point of transformer oil is around 200°C, cooking oil is around 300°C, and hello carbon 0.8 can be used till 100°C. Above these temperatures, flash point could be occurring. So oil should be used below the flash point.

2.2. Method

Water bath has been filled with water as per method described by the manufacturer. A 250 ml cylinder was filled with oil (transformer oil/ cooking oil/ Hello Carbon 0.8) up to a certain level and a PRT has been clipped into the cylinder in such a position that the insertion level of PRT (defined by the manufacture) immersed in the oil. PRT is the secondary standard device in the hierarchy of international temperature standard (ITS-90). The PRT was fixed in the cylinder so that it was held away from the sides and base of the cylinder. Two more cylinders with the same setup have been inserted in the water bath at equal distances (Fig. 1). Water bath has been set at 43.05°C.A stirrer is used to ensure the uniformity in the temperature. The temperature of the oil has been observed after allowing a sufficient time (at least 30 minutes) or after stability of temperature depends upon the condition. The bath indicates that it has reached the set-point (read out temperature) and stabilized. When the temperatures stabilize on bath and the reference indicator, value has been observed. Repeat this observation at least 3 to 5 times for observed the variation. Most occurring value may be observed if variation. Correction factor has been applied on all the measurements using conversion the error of each sensor.

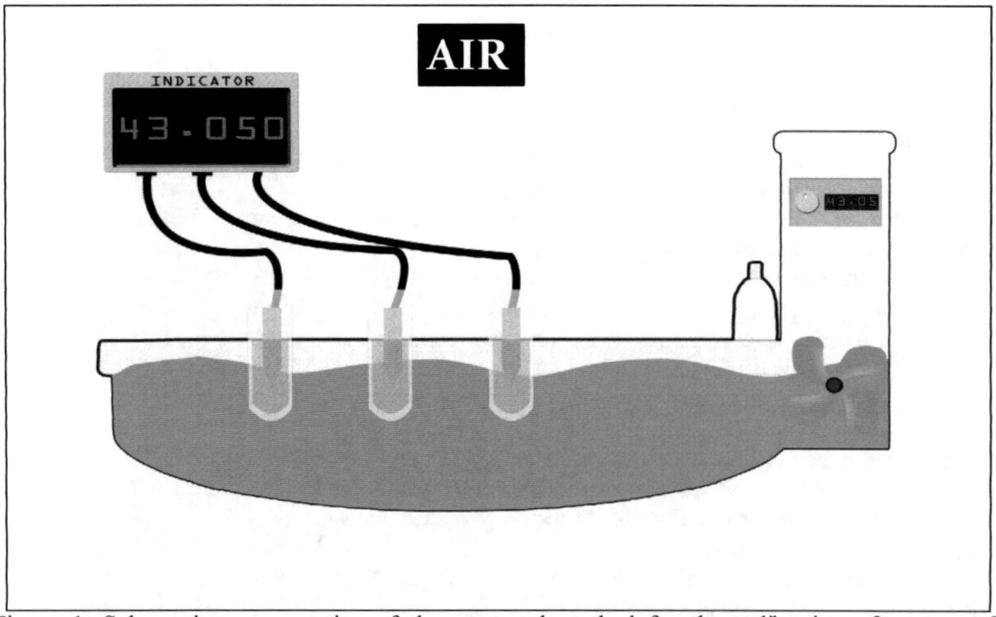

Figure 1: Schematic representation of the proposed method for the calibration of open surface water bath

3. Results & Discussion

The set temperature, readout temperature, and observed temperature of transformer oil, cooking oil, and Hello Carbon 0.8 grade, are shown in Table 1-4 respectively. Mean value and standard deviation in the results were estimated using well-known statistical formula. Generally, two types of error are estimated in the measurements. First is random error, which is always present due to operator's handling, or changes in the environmental conditions. To minimize this type of error replicate measurements were taken and their average value was calculated. Second type of error is systematic error. It is typically present due to instrumental, physical, and human limitations. This error was calculated by the formula and estimating overall (expanded) uncertainty in the measurements during calibration of water bath. All the Uncertainty measurements were performed using equation (1) to (3) as described in EURACHEM / CITAC guide.

In estimating the overall uncertainty, it is necessary to consider each source of uncertainty and treat it separately to obtain the contribution from the source. When expressed as a standard deviation, an uncertainty component is known as standard uncertainty. Equation (1) and (2) comprises of the calculation of type A standard uncertainty, and type B standard uncertainty respectively. Type A uncertainty was estimated from repeated readings and comes from statistical data (Observation). Type B uncertainty was estimated from calibration certificate of PRT and indicator, environment, readability, drift, and etc.

$$\text{Standard uncertainty for Type A} = \mu_A = \frac{s}{\sqrt{n}} \quad (1)$$

$$\text{Standard uncertainty for Type B} = \mu_B = \frac{a}{\sqrt{3}} \quad (2)$$

Where s is the standard deviation, n is the number of observation, and a is the uncertainty of the thermometer (observed value) measurement.

The total uncertainty, termed combined standard uncertainty, is an estimated standard deviation equal to the positive square root of the total variance obtained by combining all the uncertainty components (equation (3)).

$$\text{Combined uncertainty} = \mu_c = \sqrt{(\mu_A)^2 + (\mu_B)^2 + \cdots \text{etc}} \quad (3)$$

Expanded uncertainty provides an interval with in which a value of the measure and is believed to lie within a higher level of confidence. It is obtained by multiplying the combined standard uncertainty to a coverage factor (k) (equation (4)).

$$\text{Expanded uncertainty} = U = k\mu_c \quad (4)$$

k = 1, 2, and 3 for confidence level of 68 %, 95 %, and 99.7 %, respectively.

The expanded uncertainty values for transformer oil, cooking oil, and Hello Carbon 0.8 grade were calculated as 0.378°C, 0206°C, and 0.097 °C, respectively (Table 4).

4. Conclusion

PRT secondary reference device has the prospective to be utilized for the calibration of open surface water bath using oil as a secondary medium. On the basis of the experiments conducted by three different oil samples, it is concluded that the proposed method has the potential to be utilized successfully for the calibration of open surface water bath.

5. References

1. Blaivas, M., Lyon, M., Brannam, L., Duggal, S., Sierzenski, P., (2004). Water bath evaluation technique for emergency ultrasound of painful superficial structures. The American Journal of Emergency Medicine, 22(7), 589-593.

2. Veguería, SFJ., Godoy, JM., Campos, RC., Gonçalves, RA. (2013). Trace element determination in seawater by ICP-MS using online, offline and bath procedures of preconcentration and matrix eliminationTrace element determination in seawater by ICP-MS using online, offline and bath procedures of preconcentration and matrix elimination. 106, 121–128
3. Oillic, S., Lemoine, E., Gros, JB., Kondjoyan, A. (2011). Kinetic analysis of cooking losses from beef and other animal muscles heated in a water bath — Effect of sample dimensions and prior freezing and ageing. Meat Science, 88 (3). 338-346.
4. Kostka, J., Gritti, F., Guiochon, G., Kaczmarski, K. Modeling of thermal processes in very high pressure liquid chromatography for column immersed in a water bath: Application of the selected models. (2010). Journal of Chromatography A, 1217 (28), 4704-4712.
5. Ahmad, AL., Ramli, WKW, Fernando, WJW, Daud. (2012). Effect of ethanol concentration in water coagulation bath on pore geometry of PVDF membrane for Membrane Gas Absorption application in CO_2 removal. Separation and Purification Technology, 88, 11-18
6. ASTM F2473 - 12 Standard Test Method for Performance of Water-Bath Rethermalizers.
7. EURACHEM, the fitness for purpose of analytical methods (1998) (ISBN 0-948926-12-0)
8. EURACHEM / CITAC guide: Quantifying Uncertainty in analytical measurement, S. L. R. Ellison, M. Rosslein, A. Williams, 2nd edition, National Measurement System Valid Analytical Measurement (VAM) Program (2000)

Table 1: Temperature stability and uncertainty measurements in the results of Transformer oil

S. No.	Set Temp. / °C	Read Temp. / °C	Observed Temp. after Appling Correction Factor / °C			Average Temp after / °C	Error / °C	Standard deviation / °C	U_A / °C
			X1	X2	X3				
1	43.05	43.05	43.257±0.007	42.674±0.004	43.214±0.004	43.048	0.002	0.325	0.188
2	43.05	43.05	43.256±0.004	42.669±0.004	43.213±0.004	43.046	0.061	0.327	0.189

Table 2: Temperature stability and uncertainty measurements in the results of cooking oil

S. No.	Set Temp. / °C	Read Temp. / °C	Observed Temp. After Appling Correction Factor / °C			Average Temp. / °C	Error / °C	Standard deviation / °C	U_A / °C
			X1	X2	X3				
1	43.05	43.05	43.271±0.003	42.942±0.002	43.226±0.004	43.146	-0.096	0.178	0.103
2	43.05	43.05	43.270±0.004	43.000±0.003	43.215±0.003	43.162	-0.054	0.143	0.083

Table 3: Temperature stability and uncertainty measurements in the results of Hello carbon 0.8

S. No.	Set Temp. / °C	Read Temp. / °C	Observed Temp. After Appling Correction Factor / °C X1	X2	X3	Average Temp. / °C	Error / °C	Standard deviation / °C	U_A / °C
1	43.05	43.05	43.212±0.003	43.077±0.004	43.224±0.003	43.171	-0.121	0.082	0.047
2	43.05	43.05	43.210±0.004	43.078±0.002	43.223±0.003	43.170	-0.120	0.080	0.046
3	43.05	43.05	43.208±0.003	43.076±0.002	43.231±0.002	43.172	-0.122	0.084	0.048
4	43.05	43.05	43.182±0.002	43.074±0.002	43.218±0.002	43.158	-0.108	0.075	0.043

Table 4: Expanded Uncertainty Calculation

S No.	Description /°C	UA (Max.) /°C	UB /°C	UC /°C	U EXP /°C
1	Transformer Oil	0.189	0.0058	0.189	0.378
2	Cooking Oil	0.103	0.0058	0.103	0.206
3	Hello Carbon 0.8	0.048	0.0058	0.048	0.097

IOP Conf. Series: Materials Science and Engineering **51** (2013) 012016 doi:10.1088/1757-899X/51/1/012016

Real time self-mixing interferometric laser sensor for embedded applications

Usman Zabit[1], Olivier D Bernal[2,3] and Thierry Bosch[2,3]

[1] Riphah International University, Islamabad, Pakistan
[2] CNRS, LAAS, 7 avenue du colonel Roche, F-31400 Toulouse, France
[3] Univ de Toulouse, INP, LAAS, F-31400 Toulouse, France

E-mail: usman.zabit@riphah.edu.pk

Abstract. We present a Self-Mixing (SM) interferometric laser displacement sensor that is capable of providing correct target measurements in real time, even when it is subject to extraneous parasitic movements. The sensor achieves such robustness by using an embedded MEMS Solid-State Accelerometer (SSA) that has been coupled with the laser sensor. The SSA thus measures the extraneous movement acting on the laser sensor and this information is used to provide correct sensing. The proposed SSA-SM sensing system uses Consecutive-Samples based Unwrapping (CSU) to process the SM interferometric signal while a Digital Signal Processor (DSP) takes care of band-pass filtering, double integration as well as phase and gain corrections needed for the acceleration signal. Hence, a compact, real-time, precise and self-aligned SSA-SM sensor has been designed that has a displacement measurement precision of approximately 100 nm with a parasitic movement elimination of 31dB for a laser diode emitting at 785 nm.

1. Self-mixing interferometry

Self-mixing (SM) or optical feedback interferometry technique has been regularly investigated during the last two decades [1-3] for distance [4], velocity [5], vibration [6] and displacement [7] sensing applications as it results in a simple, compact, self-aligned, and low-cost sensor that becomes attractive not only for industrial use but also for large-scale consumer applications [5].

Displacement retrieval using SM interferometry remains a very active area of research [8-10] with applications as varied as run-out tracking in electric motors [9], the displacement measurement of local chest wall [11], or utilization for non-contact embedded sensing [10].

SM effect occurs in a laser when a part of the beam backscattered by a target is coupled back into the laser cavity and causes interference with the emitted beam, thus modifying the spectral properties of the laser. The variations in the optical output power of the laser diode $P(t)$ caused by this optical feedback can be written as [12]:

$$P(t) = P_0\left[1 + m.\cos\left(x_F(t)\right)\right] \tag{1}$$

where P_0 is the emitted optical power under free-running conditions, m is the modulation index and $x_F(t)$ is the laser output phase in the presence of feedback, given by:

$$x_F(t) = 2\pi \frac{D(t)}{\lambda_F(t)/2} \tag{2}$$

where $D(t)$ is the target displacement.

The emission wavelength subject to feedback $\lambda_F(t)$ is given by the phase equation:

$$x_0(t) - x_F(t) - C\sin[x_F(t) + \arctan(\alpha)] = 0 \tag{3}$$

where α is the linewidth enhancement factor, and $x_0(t)$ is the laser output phase in the absence of feedback, obtained by replacing $\lambda_F(t)$ with λ in (2), where λ is the laser diode emission wavelength under free running conditions. C is the feedback coupling factor that determines the SM operating regime and is given by [3]

$$C = \frac{\tau_D}{\tau_L}\gamma\sqrt{1+\alpha^2}\kappa_{ext} \tag{4}$$

where τ_L and τ_D are the round trip times in the internal and external cavities respectively, γ is the coupling efficiency and κ_{ext} depends linearly on the surface reflectivity of the target.

The C parameter plays an important role in the SM interferometry as variations in C cause changes in the so called SM operating regimes varying from weak to moderate to strong feedback [12], where each regime would require specific signal processing for the sake of displacement retrieval [13]. For the present work, care has been taken to maintain the SM signal in the moderate regime with C ranging from 1 to 2 as the resulting SM signal requires simplified processing and is not affected by the fringe-loss issue [14].

SM interferometry is thus a very attractive sensing scheme in the way that it allows a simple laser diode (LD) package (containing its built-in photodiode) to be used as a micro-interferometer, a laser source as well as a detector. This thus allows a compact, miniaturized, low-cost and self-aligned sensor capable of nanometric measurement precision.

Now, it needs to be said that SM displacement sensors have traditionally needed a stationary support (such as an optical table and/or anti-vibration material) to guarantee an accurate target measurement as any parasitic/extraneous movement of the SM sensing laser system can corrupt the true target displacement measurement. Such a situation has thus restricted the use of SM sensors for embedded and/or industrial applications where hostile environment can introduce parasitic extraneous movements.

So, keeping in mind such a limitation that can potentially limit the use of such optical sensors, a Solid-State Accelerometer (SSA) coupled Self-Mixing (SSA-SM) sensing system has been recently proposed [15] and evaluated [10] that measures true target displacement even if the laser sensor is itself subject to a displacement (noted as Ds). A schematic diagram of such a SSA-SM sensing system is presented in figure 1. The solid-state MEMS (micro-electro-mechanical system) accelerometer measures the laser sensor displacement and a subsequent subtraction of this displacement from the global (corrupted) SM displacement provides the true target displacement.

In this paper, after a presentation of the signal processing needed for a SSA-SM sensing system, we shall detail the approach taken for the design of a real-time SSA-SM sensing system. An all digital signal processing approach has been developed that takes care of the band-pass filtering, double integration as well as phase and gain corrections required by the acceleration signal. The presented real-time SSA-SM sensor system is based on a LD package from Sanyo (DL7140) emitting at λ=785nm with an output power of 60mW. The accelerometer LIS344ALH is from ST® with a typical noise resolution of 50μg/√Hz and a full-scale range of ±2g. Finally, the results based on this real-time sensing prototype would be presented and compared with a commercial PZT (Piezo-electric Transducer) from Physik Instrumente (P753.2CD) equipped with a built-in capacitive feedback sensor (CFS) of 2nm resolution. The PZT has been used as the target in our set-up (see figure 1) and the CFS has served as a reference sensor.

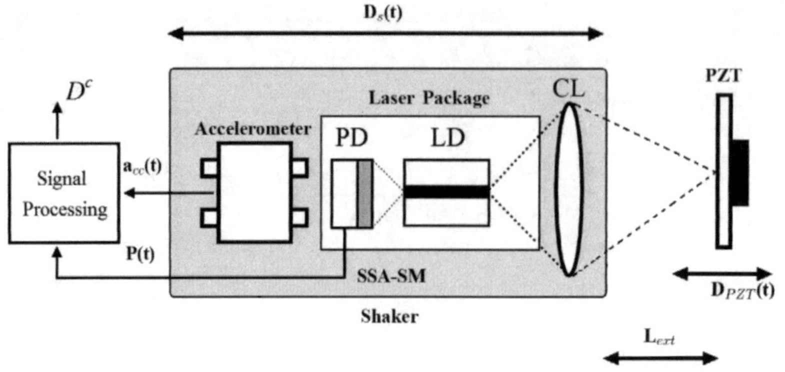

Figure 5. Schematic block diagram of the Solid-State Accelerometer (SSA) coupled Self-Mixing (SSA-SM) sensing system: photodiode (PD), laser diode (LD), collimating lens (CL), and piezoelectric transducer (PZT).

Let us now start with a presentation of the signal processing needed for the real-time SSA-SM sensing system capable of removing the influence of extraneous movement acting on it.

2. Signal processing

The signal processing steps considered necessary for a correct working of our SSA-SM sensing system are presented in figure 2. Let us now present each of the three main processing blocks as detailed below.

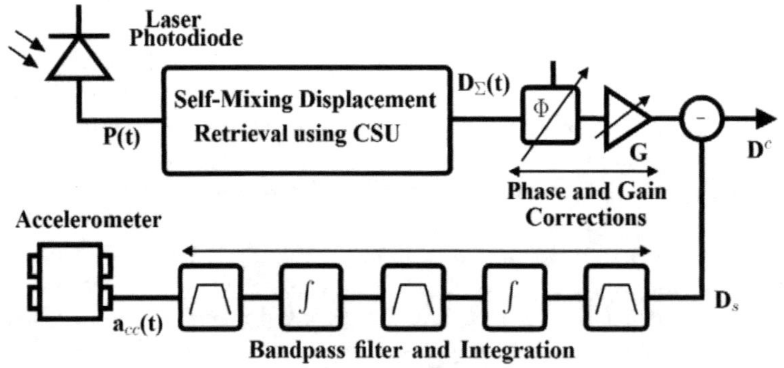

Figure 2. Schematic block diagram of the signal processing steps needed for the real-time Self-Mixing interferometric laser displacement sensor with embedded MEMS accelerometer.

2.1. SM Laser Interferometric Signal

The variations in the optical output power of the laser diode P(t) are monitored through the built-in photodiode (PD) contained in the LD package. For real-time SSA-SM sensing system, SM displacement retrieval has been done by using the Consecutive Samples based Unwrapping (CSU) method [16]. This displacement retrieved from the signal P(t) has been denoted here as D_Σ since it is the sum of the true target displacement DPZT and the displacement undergone by the sensor D_s.

The working of CSU method is shown in figure 3. It starts by retrieving the SM based displacement through fringe-counting resulting in an initial precision of $\lambda/2$. Then, in a manner similar to the fully analog SM processing method [17], it recovers the target displacement by adding the fringe-counting based staircase signal to the normalized SM signal. So, it is able to retrieve SM based displacement with a precision of approximately $\lambda/8$.

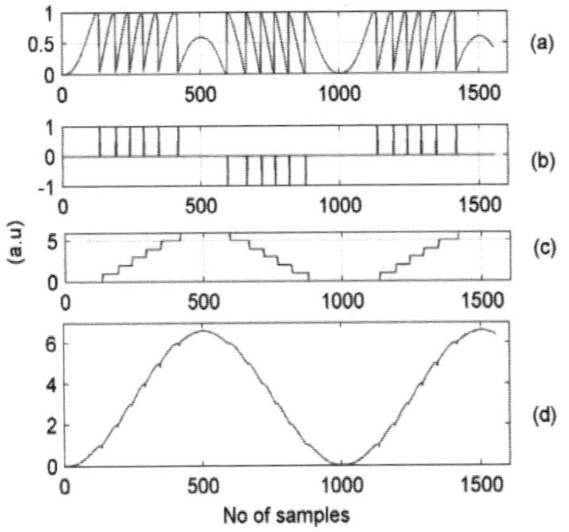

Figure 3. Displacement retrieval through Consecutive-Samples based Unwrapping [15]. (a) normalized SM signal corresponding to C=1.5, (b) fringe discontinuity detection, (c) staircase signal representing the fringe count and (d) the retrieved signal proportional to target excitation signal obtained by the addition of signals (a) and (c).

It may be said that in order to process the SM signal in real-time, the fully analog SM processing technique proposed by Norgia et al [17] may have been considered. However, keeping in mind the non-harmonic nature of D_Σ, it is important to use an approach capable of processing non-harmonic movements without suffering from drift effects. Hence, the SM signal has been processed in real-time using the above-mentioned CSU method.

It may also be added that the SM sensing precision of direct unwrapping approach is lesser than a segmented unwrapping approach, as highlighted by [18]. Recent work has shown as to why such a segmented processing improves the sensing precision [19]. However, this segmented unwrapping approach [18-19] has not been used due to increased processing complexity hindering a real-time sensor design, a condition needed for the present sensor.

For the actual SSA-SM sensing system, an integrated micro-converter AduC7020 from Analog Devices® working at 40 MIPS has been used that has integrated Analog-to-Digital Converter (ADC) and Digital-to-Analog Converter (DAC) modules. The incoming signal P(t) is processed into the output displacement signal D(t) at 125KHz with an approximately $\lambda/8$ precision. So, the SM based displacement D_Σ is updated every 8µs.

As far as the maximum measurable target speed using this prototype is concerned, the experimental data has shown that 20 samples per SM fringe are normally needed in order to correctly detect the fringe-discontinuity of each SM fringe having its associated noise. This then provides the maximum measurable target speed of 2.45mm/s for the real-time SM signal based displacement retrieval method with the above-mentioned configuration [16].

2.2. Accelerometer Signal
The displacement undergone by the sensor itself (noted here as D_s) is recovered by processing the acceleration signal provided by the SSA. In order to do so, the acceleration signal $a_{cc}(t)$ is filtered and integrated twice, as indicated in the lower half of figure 2. The band-pass filtering is done over the operating bandwidth of 40Hz-500Hz so that low-frequency drifts that can falsify the subsequent integration steps as well as high frequency acceleration signal saturation can both be avoided.

The choice of a given low cut-off frequency is closely related to the overall SSA-SM sensing system measurement precision. As displacement extraction from acceleration signal involves double integration, so the noise power spectrum density of the displacement S_D can be given in the frequency domain as [15]

$$S_D(s) = \frac{S_{acc}(s)}{s^4} \tag{5}$$

Consequently, the displacement resolution of the SSA signal depends on the low cutoff frequency of the system. For a 20Hz lower cutoff, LIS344ALH is thus expected to introduce an RMS error of approximately 40nm.

For the actual SSA-SM sensing system, the integration and filtering stages (indicated in figure 2) needed for extracting the displacement D_s from $a_{cc}(t)$ signal provided by LIS344ALH have been implemented on an integrated Digital Signal Processor (DSP) from Microchip®, called the dsPIC33FJ128GP. It has also integrated ADC and DAC modules which allow a simple interfacing. The incoming signal $a_{cc}(t)$ is processed into the displacement signal $D_s(t)$ at 10KHz. Such a sampling rate is sufficiently high to correctly process the acceleration signal of 40Hz-500Hz operating bandwidth.

2.3. Matching of SM signal with Acceleration signal

In order to match the two signals (D_Σ and D_s), phase (ϕ) and gain (G) corrections need to be done. This is necessary as both signals i.e. the self-mixing signal and the acceleration signal pass through their respective signal amplification, acquisition, and processing stages all of which may introduce differences in phase and gain between the two signals.

So, in order to measure the phase and gain coefficients between the two displacements signals, the SSA-SM sensor was subjected to vibrations within the 40Hz-500Hz bandwidth. This allowed us to recover the coefficients that have been presented in figure 4.

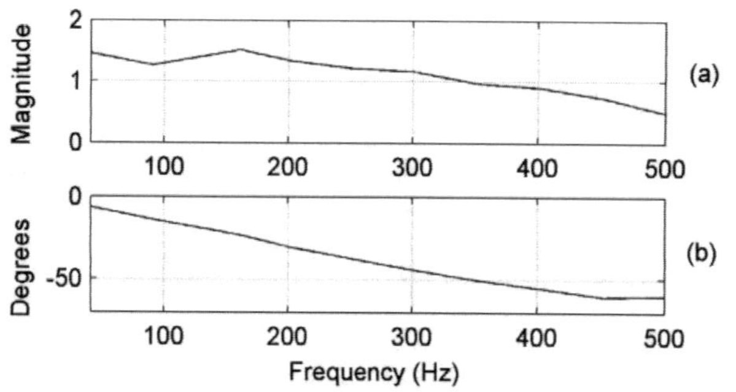

Figure 4. (a) Gain transfer function and (b) Phase transfer function between SM based displacement D_Σ and acceleration based displacement D_s. A digital filter having the same Bode diagram has been implemented on the DSP.

Then, a custom filter was designed based on these coefficients and it was also implemented on our DSP. Such a filter thus allowed us to match D_Σ with D_s (see figure 2).

Finally, after this phase and gain matching of the two signals (D_Σ and D_s), we need only perform the last step of their subtraction that allows us to obtain the corrected displacement D^c. As the signals D_Σ and $a_{cc}(t)$ are both processed by the DSP, so the subtraction is also done internally by the DSP and final corrected output D^c is hence provided by the DAC of our DSP. Such a solution thus allows us to recover the corrected displacement in real-time.

3. Experimental results

Let us now look into some of the experimental results for the SSA-SM sensing system. As indicated in figure 1, the SSA-SM sensor was mounted on a mechanical shaker that has been used to generate extraneous movement acting on the SSA-SM sensor. The sensor measured the target displacement (we have used the PZT as target while the CFS has been used as a reference sensor of the target motion) while being subject to these extraneous movements.

IOP Conf. Series: Materials Science and Engineering **51** (2013) 012016 doi:10.1088/1757-899X/51/1/012016

Initially, in order to validate if the SSA-SM sensing system was able to eliminate the influence of extraneous parasitic movements, it was decided to keep the target stationary while the shaker was excited at different frequencies within the 40Hz-500Hz operating bandwidth. It was thus measured that the real-time SSA-SM sensing system eliminated these parasitic vibrations to an average RMS value of 97% or a reduction of 31dB over the operating bandwidth of 40Hz-500Hz. The measurement precision was found to be approximately 100nm.

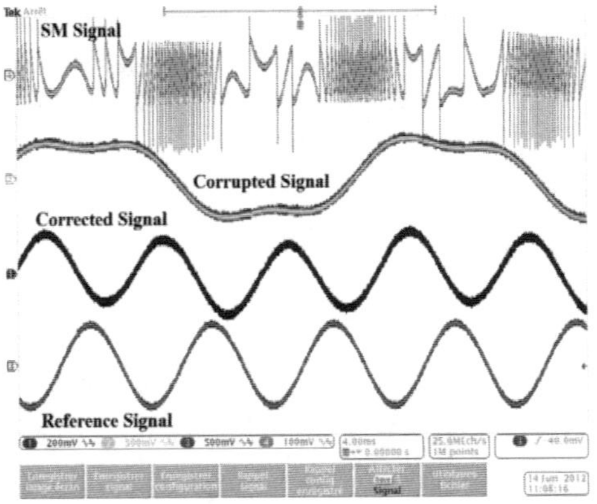

Figure 5. Experimental real-time sinusoidal target displacement measurement in the presence of an extraneous sinusoidal movement at 40Hz while the target vibrates at 117Hz: (green) SM signal, (light blue) corrupted movement, (dark blue) corrected movement, and (purple) reference sensor movement.

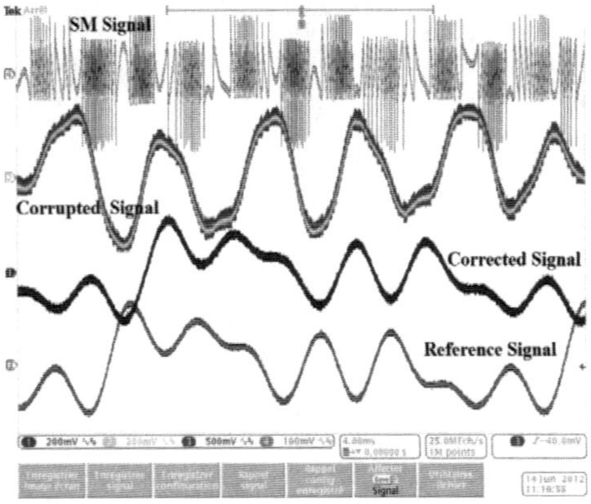

Figure 6. Experimental real-time correction of an arbitrary target movement corrupted by a different arbitrary movement acting on the SSA-SM sensor: (green) SM signal, (light blue) corrupted movement, (dark blue) corrected movement, and (purple) reference sensor movement.

Subsequently, after this validation, different tests were conducted to verify a correct working of our real-time SSA-SM system for the sake of target displacement measurements. For example, both the shaker and the PZT were excited with either a harmonic or an arbitrary excitation signal.

Figure 5 presents the correction result for the case where the target i.e. the PZT was made to vibrate at 117Hz with an amplitude of 4μm while the extraneous harmonic movement acting on the SSA-SM sensor (provided by the shaker) was at 40Hz. It can be seen in figure 5 (light blue curve) that the corrupted signal i.e. D_{Σ} is not at all indicative of true target movement (seen in figure 5 purple curve). However, a correct target displacement measurement has been made by our real-time SSA-SM sensing system as indicated by the dark blue curve of figure 5. A close resemblance between the corrected signal and the reference signal is thus seen. The observed phase difference between the corrected and the reference signal is due to the phase difference between the PZT and the CFS and is of no consequence for our sensing application.

Another real-time displacement measurement result is presented in figure 6 where the shaker and the target were both excited by different arbitrarily shaped excitation. The shaker was thus excited with a signal composed of 140Hz-280Hz while the PZT was excited with a signal composed of 31Hz-62Hz-124Hz-155Hz. As a consequence, the displacement D_{Σ} measured by the SM signal is completely corrupted (figure 6 light blue curve) and is not at all indicative of true target movement (figure 6 purple curve). However, the influence of the extraneous movement disturbing the SSA-SM sensing system has been correctly eliminated by our real-time signal processing which allowed a correct target movement measurement as seen in figure 6 dark blue curve.

Other tests were also conducted that also verified a correct working of our real-time SSA-SM sensing system. figure 7 depicts the case where the target vibration as well as the extraneous vibration are quite close to each other in terms of frequency (79Hz for the target while 83Hz for the extraneous movement). Such a case results in a corrupted signal that appears to be amplitude modulated as seen in figure 7 light blue curve. However, a correct movement has again been retrieved as seen in figure 7 purple curve.

Another interesting case with spectral overlap is presented in figure 8. Here, a test was conducted where the shaker and the PZT were both excited at the same frequency (in this case, 84Hz). Correct results were again obtained in spite of the fact that the target and the extraneous motion have exactly the same frequency. Such a test is important as it validates the working of our sensor for such embedded sensing applications where a mechanical coupling between the target and the sensor may cause the sensor to be subject to extraneous motion at exactly the same frequency as that of the target.

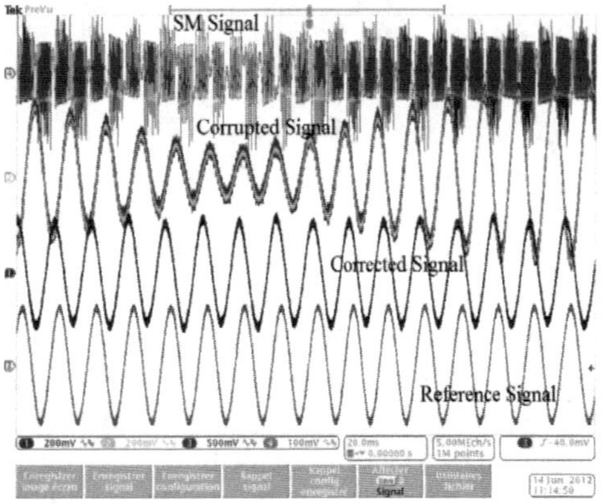

Figure 7. Experimental real-time sinusoidal target displacement measurement in the presence of an extraneous sinusoidal movement at 83Hz while the target vibrates at 79Hz: (green) SM signal, (light blue) corrupted movement, (dark blue) corrected movement, and (purple) reference sensor movement.

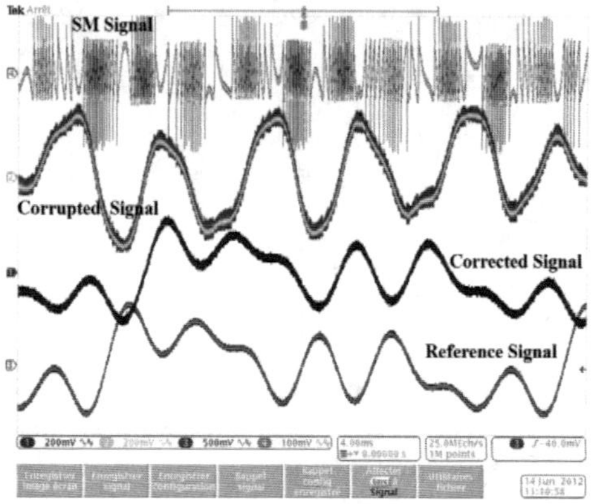

Figure 8. Experimental real-time sinusoidal target displacement measurement in the presence of an extraneous sinusoidal movement at 84Hz while the target also vibrates at 84Hz: (green) SM signal, (light blue) corrupted movement, (dark blue) corrected movement, and (purple) reference sensor movement.

4. Conclusion

In this paper, a real-time implementation of a Self-Mixing (SM) interferometric laser diode (LD) based displacement sensor coupled with a Solid-State Accelerometer (SSA) has been presented. The said real-time SM displacement sensor system is based on a LD package from Sanyo (DL7140) emitting at λ=785 nm with an output power of 60 mW. The accelerometer LIS344ALH is from ST® with a typical noise resolution of 50μg/√Hz. Real-time signal processing has been achieved by an embedded micro-converter and a DSP.

Such an SSA-SM sensor allows measuring correct target movements even when the LD based SM sensor is subject to extraneous movements. This results in a displacement/vibration sensing system that can be used for embedded applications as there is no more need of keeping the sensor stationary. Such an approach opens the way for the use of such laser sensors in conditions where the use of anti-vibration support is not available or possible. The proposed DSP based data fusion between the acceleration signal and the SM interferometric signal enables a robust, compact, real-time and low-cost sensing system.

Acknowledgment
Usman Zabit warmly acknowledges all his former colleagues working alongside him on the embedded laser displacement sensor project at the CNRS, LAAS, 7 avenue du colonel Roche, F-31400 Toulouse, France as well as the Univ de Toulouse, INP, LAAS, F-31400 Toulouse, France.

References

[1] Donati S 2012 Developing self-mixing interferometry for instrumentation and measurements *Laser & Photon. Rev.* **6** 393417.
[2] Giuliani G and Donati 2005 S Laser interferometry *Unlocking Dynamical Diversity: Optical Feedback Effects on Semiconductor Lasers* ed D M Kane and K A Shore (Chichester: John Wiley & Sons Ltd) pp. 217-256.
[3] Bosch T, Bes C, Scalise L and Plantier G 2006 Optical feedback interferometry *Encyclopedia of Sensors* American Scientific Publishers **7** pp. 107–127.
[4] Mourat G, Servagent N and Bosch T 2000 Distance measurement using the self-mixing effect in a three-electrode distributed Bragg reflector laser diode *Opt. Eng.* **39** 738
[5] van der Lee A, Carpaij M, Moench H, Schemmann M and Pruijmboom A, "A miniaturized VCSEL based sensor platform for velocity measurement", Proc. of the 25th IEEE International Instrumentation and Measurement Technology Conference, pp 141-143, 2008.
[6] Giuliani G, Bozzi-Pietra S and Donati S 2003 Self-mixing laser diode vibrometer *Meas. Sci. Technol.* **14** 24-32
[7] Ottonelli S, De Lucia F, di Vietro M, Dabbicco M, Scamarcio G and Mezzapesa F P 2008 A Compact Three Degrees-of-Freedom Motion Sensor Based on the Laser-Self-Mixing Effect *Photonics Technology Letters, IEEE* **20** 1360-2
[8] Zabit U, Bernal O D and Bosch T 2013 Self-mixing laser sensor for large displacements: signal recovery in the presence of speckle *IEEE Sensors Journal* **13** 824-31
[9] Atashkhooei R, Urresty J-C, Royo S, Riba J-R and Romeral L 2013 Runout Tracking in Electric Motors Using Self-Mixing Interferometry *Mechatronics, IEEE/ASME Trans.* doi: 10.1109/TMECH.2012.2226739,.
[10] Zabit U, Bernal O D and Bosch T 2013 Design and analysis of an embedded accelerometer coupled Self-Mixing laser displacement sensor *IEEE Sensors Journal* **13** 2200-7.
[11] Milesi I, Norgia M, Pompilio P P, Svelto C and Dellaca R L 2011 Measurement of Local Chest Wall Displacement by a Custom Self-Mixing Laser Interferometer *Instrumentation and Measurement, IEEE Trans.* **60** 2894-901
[12] Plantier G, Bes C and Bosch T 2005 Behavioral model of a self-mixing laser diode sensor *Quantum Electronics, IEEE Journ.* **41** 1157-67
[13] Zabit U, Bosch T, and Bony F 2009 Adaptive transition detection algorithm for a self-mixing displacement sensor *IEEE Sensors Journ.* **9** 1879–86
[14] Zabit U, Bosch T, Bony F and Rakic A D 2010 A self-mixing displacement sensor with fringe-loss compensation for harmonic vibrations *IEEE Photonics Technology Letters* **22** 410–2
[15] Zabit U, Bernal O D, Bosch T and Bony F 2011 MEMS accelerometer embedded in a self-mixing displacement sensor for parasitic vibration compensation *Optics Letters* **36** 612-4
[16] Zabit U, Bernal O D and Bosch T 2012 Self-mixing sensor for real-time measurement of harmonic and arbitrary displacements *Inst. and Meas. Techn. Conf. (I2MTC), 2012 IEEE* 13-16
[17] Norgia M and Pesatori A 2011 Fully analog self-mixing laser vibrometer *Inst. and Meas. Techn. Conf. (I2MTC), 2011 IEEE*, pp.1-4
[18] Y. Fan, Y. Yu, J. Xi, and J. F. Chicharo 2011 Improving the measurement performance for a self-mixing interferometry-based displacement sensing system *Applied Optics* **50** 5064–72
[19] Zabit U, Bernal O D and Bosch T 2013 Study of Laser Feedback Phase under Self-Mixing leading to Improved Phase Unwrapping for Vibration Sensing *IEEE Sensors Journ.* doi:10.1109/JSEN.2013.2276106

An Online Risk Monitor System (ORMS) to Increase Safety and Security Levels in Industry

M Zubair[1,2], Khalil Ur Rahman[1], Mehmood Ul Hassan[3]
[1]College of Engineering, *Department of Nuclear Engineering: Kyung Hee University*, Yongin Si, Gyeonggi Do, 446-701, Republic of Korea
[2]Department of Basic Sciences, University of Engineering and Technology, Taxila, Pakistan
Department of Agronomy, Pir Mehr Ali Shah, Arid Agriculture University, Rawalpindi, 46300, Pakistan

E-mail: zubairheu@gmail.com

Abstract. The main idea of this research is to develop an Online Risk Monitor System (ORMS) based on Living Probabilistic Safety Assessment (LPSA). The article highlights the essential features and functions of ORMS. The basic models and modules such as, Reliability Data Update Model (RDUM), running time update, redundant system unavailability update, Engineered Safety Features (ESF) unavailability update and general system update have been described in this study. ORMS not only provides quantitative analysis but also highlights qualitative aspects of risk measures. ORMS is capable of automatically updating the online risk models and reliability parameters of equipment. ORMS can support in the decision making process of operators and managers in Nuclear Power Plants.

1. Introduction

Over the past years, many nuclear power plant (NPP) organizations have performed probabilistic safety assessments (PSAs) to enhance the safety level of NPP. These PSA studies is an effective tool because it assist plant management to get more benefits for plant safety but any PSA used to support decision making must have a defensible basis therefore it is very important that regulatory body accept Living PSA. LPSA provides basis for risk informed approach to decision making.
A risk monitor is mainly used to calculate instantaneous risk [1]. The first risk monitors were put into operation in 1988. The number of risk monitors worldwide has increased to over 150. The risk monitors are used for quantitative analysis as well as qualitative analysis such as safety function, safety system. There are different types of risk measures like;

- Baseline risk which is the numerical value of the risk (CDF, LERF, etc.) calculated by the PSA with all components available to carry out their safety function.
- The average risk which is normally calculated by the Living PSA for full power operation, Average risk is calculated when average maintenance unavailability Introduced and it is always greater than the baseline risk.
- The point-in-time risk is the level of risk is related to a specific plant. The point-in-time risk will change as the plant configuration and environmental factors change as shown in figure.1 [2].

[1,2] zubairheu@gmail.com

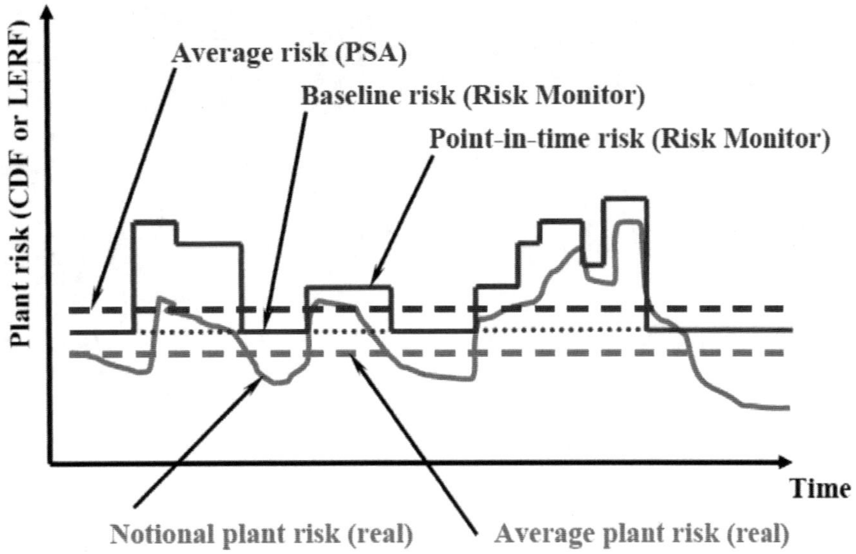

Figure.1 Average, baseline and point in time risk

The ORMS presented in this article capable to calculate changes in configuration and reliability of components in NPP. ORMS is based on full power, internal event Level 1 PSA and update risk models regularly and automatically.

2. Structure of Online Risk Monitor System (ORMS)

The basic structure of ORMS consists of five modules known as;

- Reliability Data Update Module (RDUM)
- Running Time Update
- Redundant System unavailability Update
- Engineered Safety Function (ESF) Unavailability Update
- General System Update

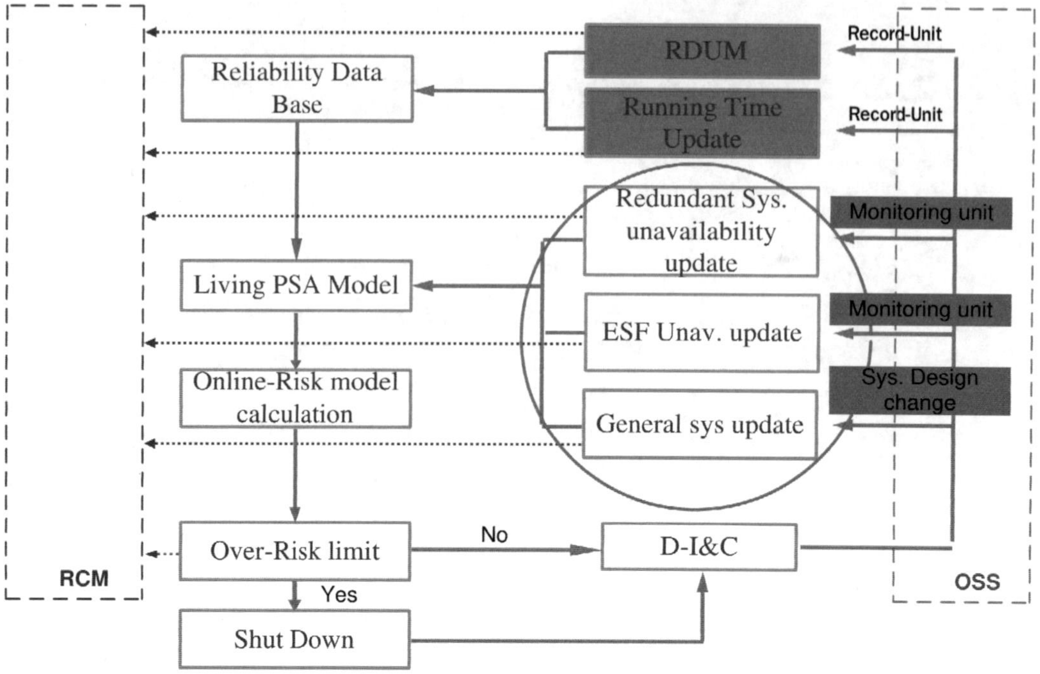

Figure.2 Structure of ORMS

These five modules are shown in Figure.2. The first two modules RDUM and Running Time Update receive information from D- I & C (Digital Instrumentation and Control) system, analysed data quantitatively and supply feedback to Reliability Data Base. The remaining three modules receive information from monitoring unit & system design change unit and analysed data qualitatively. The qualitative and quantitative output of these three modules in combination with reliability data base module is provided to living PSA model respectively. After getting information the online risk model makes a quick calculation of following factors;

- Core damage frequency
- Importance factor
- Allowed configuration time
- Qualitative risk information

In view of calculation, the online risk model makes it possible to shut down plant if risk exceed over a limit and continue updating process if risk levels liaise within limits.
To prevent failures Reliability Centered Maintenance (RCM) gathered and compares all updated data for analysis [3]. The purpose of comparison in RCM is to identify needed changes in the existing program and thereby optimize the facility's Preventive Maintenance program.

3. Specification of Modules in ORMS

3.1. RDUM and Running Time Update

The RDUM work in such a way that it uses Bayes' Theorem and combination of different distributions for the calculation and updating of parameters, Figure. 3 describe this concept clearly. Two types of distributions have been used.
(1) Beta distribution with binomial likelihood function for the calculation of demand failure probability. Equation 1 and 2 explain the key results of these distributions and the calculation process or steps can be seen from literature [4].

$$\alpha_{\mathrm{post}} = k + \alpha_{prior} \quad (1)$$

$$\beta_{post} = \mathrm{n} - \mathrm{k} + \beta_{prior} \quad (2)$$

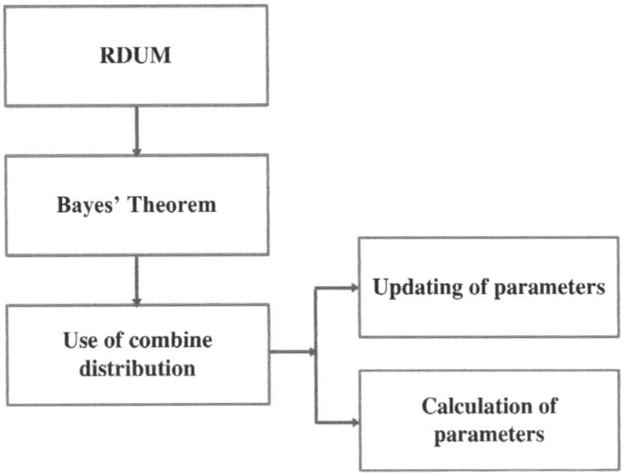

Figure.3 Function of RDUM

(2) Gamma distribution with Poisson likelihood function to update running time, equation (3) and (4) explain final results.

$$\alpha_{\mathrm{post}} = \mathrm{X} + \alpha_{prior} \quad (3)$$

$$\beta_{post} = \mathrm{t} + \beta_{prior} \quad (4)$$

3.2. Redundant, ESF and general system Unavailability update

The function of these three modules is to make qualitative analysis of data and provide this information to LPSA model. The redundant and ESF modules receive information from monitoring unit, while general system module updated as changes occur in system design. The logical configuration of monitoring system is shown in figure.4

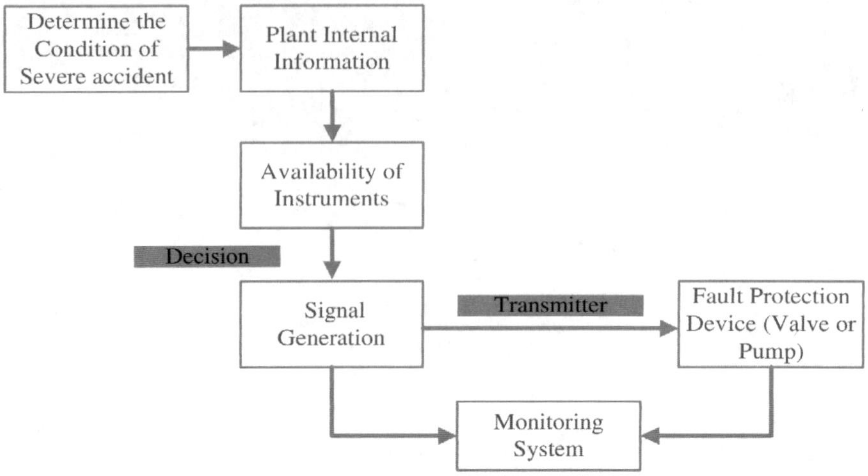

Figure.4 Logical Configuration of monitoring system

There are three main steps in figure.4.

(i) Signal generation process which includes human operator's judgment and proper action (push button), normal operation of command generation equipment

(ii) Success of command transmission through electrical wire or pressure sensing line

(iii) Normal response of the actuating device to a given command. The Fault tree analysis is made to find the unreliability (failure probability) in the control command generation process.

The system unavailability increase risk level. Let us consider that R_1 represent enlarged value of risk measure in Core Damage Frequency (CDF) when component is not performing its function. [5]. R_0 is the reduced CDF when component is in operation (means component available). In terms of R_1 and R_0 the increase ΔR in risk level associated with the Allowed Outage Time (AOT) then;

$$\Delta R = R_1 - R_0$$

Using the above expression, Allowed Outage Time can be calculated as below;

r = Allowed Outage Time single-event

$= (R_1 - R_0)*d$

And

Ry = AOT risk calculated on annul basis

$= f.r$

$= f. (R_1 - R_0)*d$

R_1 and R_0 obtained from PSA data.

4. ORMS Graphical Interphase

The graphical interface of ORMS enables a user to calculate and update data within a few minutes. The login screen is shown in figure. 5.

Figure.5 Login screen of ORMS

After providing user name and pass ward the main page of ORMS open as shown in figure.6

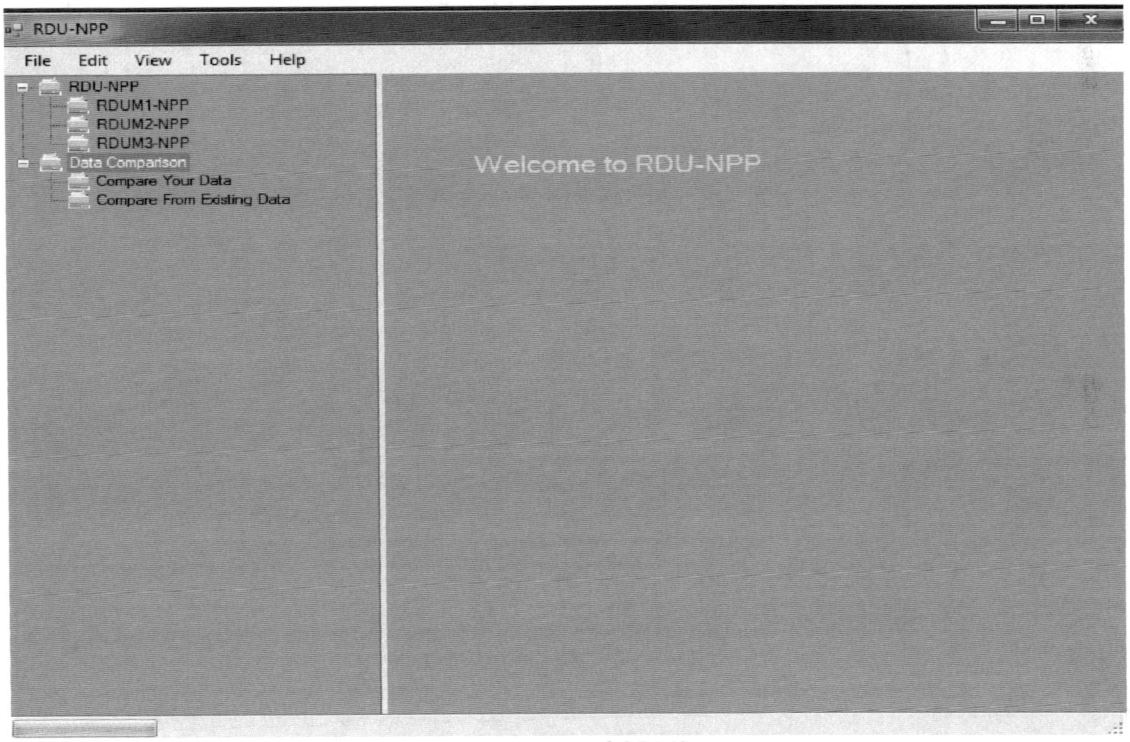

Figure.6 Main page of ORMS

Now at this stage if user wants to calculate operational failure rate than after assign the values of number of failures and number of demands, RDUM-1 will provide required results and if it is needed to update running time than RDUM-2 execute updating process, these steps are shown in figure. 7

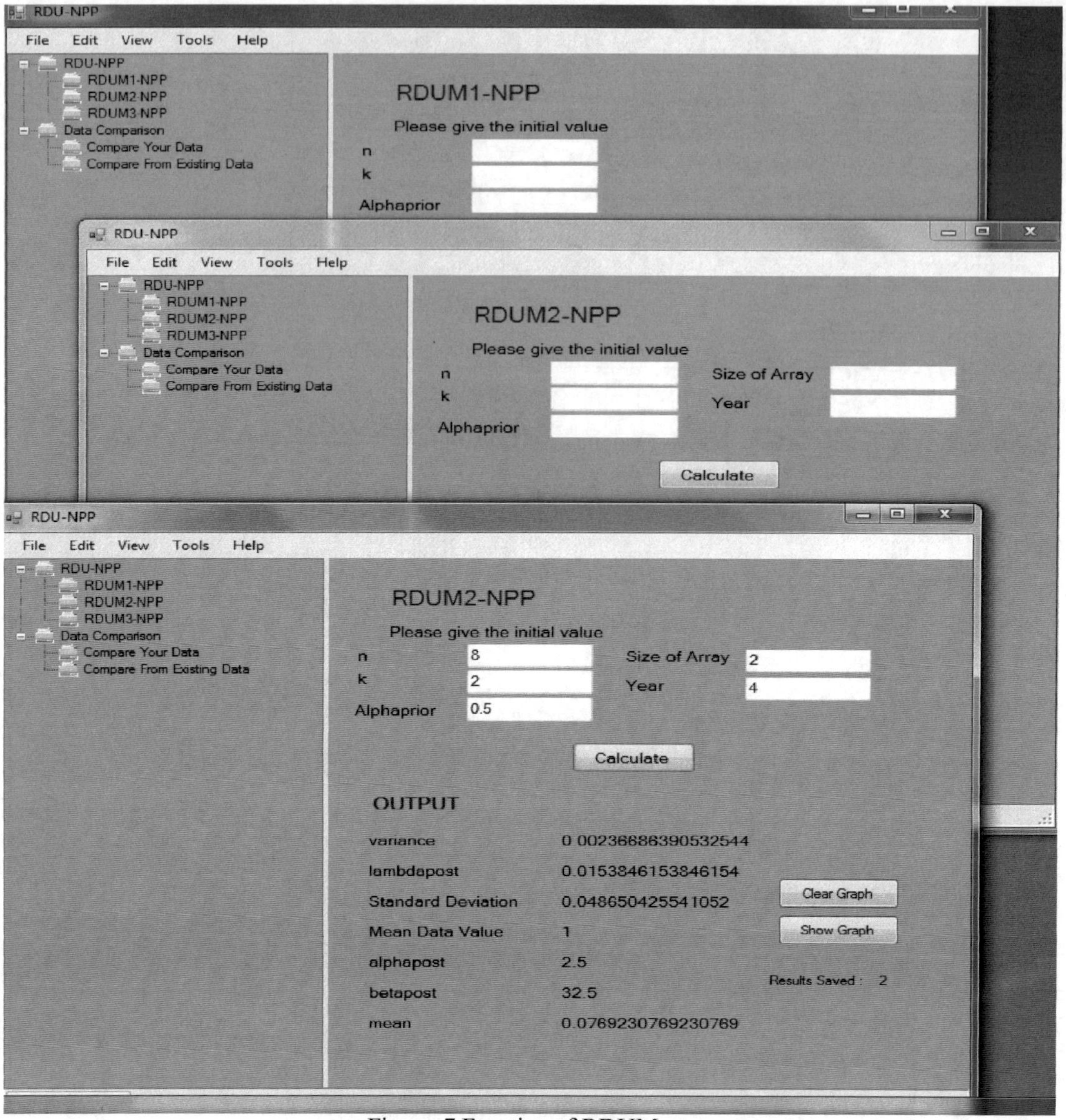

Figure. 7 Function of RDUM

5. Conclusion

To achieve safety standards the utilization of risk monitors in decision making process seems obvious. In this paper Online Risk Monitor System (ORMS) has been developed to measure qualitative and quantitative aspects of data. Five modules of ORMS has been explain which are capable of updating Reliability data base, LPSA model and risk model calculation. With the help of ORMS the operator can update PSA model to online risk monitor model. In literature limited data is available for LPSA, so presented work also provide a useful guide line for new researcher to understand the concept of LPSA and risk monitoring.

References

[1] IAEA, 1999 *TECDOC-1106*
[2] NEA/CSNI/R 1996 *NEA/ CSNI/R* **2** 1–31
[3] IAEA, 2007 International Atomic Energy Agency *TECDOC-1590*
[4] Zubair M, Zhijian Z 2011 *Annals of Nuclear Energy* **38** 2575-80
[5] NURGE/CR-6141 1994 Hand book of methods for risk-based analysis of technical specifications

FPGA based Smart Wireless MIMO Control System

Syed M Usman Ali[1], Sajid Hussain [2], Ali Akber Siddiqui [3], Jawad Ali Arshad [4], Anjum darakhshan [5]
Department of Electronic Engineering, NED University of Engineering & Technology Karachi, Usman Institute of Technology, Hamdard University Karachi, Institute of Industrial Electronics Engineering, Karachi, Sir Syed University of Engineering and Technology, Karachi

Email: uashah68@neduet.edu.pk, sajid@iiee.edu.pk, ali124k@hotmail.com, anjum_iiee@yahoo.com, jawwadaliarshad@gmail.com.

Abstract: In our present work, we have successfully designed, and developed an FPGA based smart wireless MIMO (Multiple Input & Multiple Output) system capable of controlling multiple industrial process parameters such as temperature, pressure, stress and vibration etc. To achieve this task we have used Xilinx Spartan 3E FPGA (Field Programmable Gate Array) instead of conventional microcontrollers. By employing FPGA kit to PC via RF transceivers which has a working range of about 100 meters. The developed smart system is capable of performing the control task assigned to it successfully. We have also provided a provision to our proposed system that can be accessed for monitoring and control through the web and GSM as well. Our proposed system can be equally applied to all the hazardous and rugged industrial environments where a conventional system cannot work effectively.

1. Introduction

Technology is evolving on daily bases and as it evolves, the manpower reduces and the total burden fall on machines to monitor the different parameters continuously or else system failure is eminent. Safeties of plant machineries and as well as plant workers are very important matters. For example, the temperature which is the most widely monitoring parameter in the various parts of industrial processes and machines in order to obtain the quality products from these machines utilized for such tasks. If we required to monitor the underground temperature of geothermal power plant or the radiation levels in the hazardous process and some machineries installed in the
basements where to keep the human presence around the clock is difficult, then for such situations we required a control system capable to monitor the temperature and other process parameters of such environments through wireless control. For such specific situations we developed a wireless smart MIMO system capable of monitoring multiple process parameters simultaneously with control actuation abilities. We have successfully developed this system by utilizing an FPGA that is capable of achieving the tasks according to our requirements.

The purpose of using FPGA instead of a conventional microcontroller is that we need to monitor multiple parameters and to control these parameters as well. For such purpose we required a specific controllers /devices that are capable of parallel processing and since the conventional microcontrollers do not support this feature of parallel processing.

2. Controllers/Devices Used

2.1.1. FPGA (Field Programing Gate Array)

FPGA is widely used in emerging control automation systems due to its promising features as controller. In the present work, we are using Xilinx XC3S500E FPGA board as shown in figure (1) developed by DIGILENT [1]. It has over 500,000 gates, Two DB9 RS-232 connectors, 64MB Micron DDR SDRAM, 16MB Numonyx Strata Flash, and 2MB ST Microelectronics Serial Flash.

Fig. (1) Xilinx XC3S500E FPGA [1]

2.1.2. GSM (SIM300)

GSM (Global system for mobile communication) is a system capable of communicating wirelessly with anyone around the globe. GSM module SIM300 is used as shown in figure (2) in our proposed system. The GSM/GPRS works on frequencies EGSM 900 MHz, DCS 1800 MHz, and PCS1900 MHz, SIM300 also provided the support of GPRS coding schemes CS-1, CS-2, CS-3 and CS-4[2]. In our system we used GSM for communicating the acquired controlled information from the automated plant/system to concerned people wirelessly; if the system detects an undesirable reading then we have also introduced a provision to control such parameters via messaging if required.

Fig. (2) SIM300 GSM Modem

3. Relevant Research

Prof. S. S. Sarade, Prof A. C. Joshi, Prof. Sachin S. Patel, and Prof. A.N. Shinda, proposed a Wireless Temperature Monitoring system based on WSN (Wireless Sensor Network). They designed a WSN based on PIC16F877 microcontroller. There are several individual devices known as sensor nodes and each sensor node consists of TCN-75 which is a 2 wire serial temperature sensor and Thermal monitor that is interfaced to PIC16F877, and PIC is communicating with the host system via Zigbee 802.15.4 protocol. These sensor nodes are used to inform about various parameters such as temperature, sound, motion and pressure etc. [3]. Theophilus Wellem and Bhudi Setiawan have also proposed a sensing and controlling system based on ATMega8535. The sensor is continuously monitoring the temperature of the room and informing the concerned member via message using GSM modem [4]. Bing and Mwenyao also proposed a system based on Zigbee module which is wireless temperature monitoring and control system for communication room. They used JN5121 Zigbee wireless microcontroller and Sht11 Temperature sensor [5]. Manoranjan Das, and Banoj Kumar Pand presented an idea of a wireless sensor node using the field programmable gate array (FPGA) based architecture for an early detection of hazards (e.g fire and gas-leak) in mines area[8]. SUN Yu-jia,WANG Xiao-ming,JIA Fang-xiu, and YU Ji-yan introduced an idea to implement the WSN system based on Cortex-M3 Microcontroller STM32F103RE chip [9].

In comparison to our proposed system, we have chosen FPGA for controlling the multiple inputs and output devices in parallel processing configuration and interfaced with the central monitoring computer. The basic reason for choosing FPGA instead of a conventional microcontroller is to make our processing faster by utilizing the parallel processing capabilities of FPGA. Web based database application is also developed and designed to monitor the parameters online; moreover, we have also equipped our system with GSM module as well. GSM is used for informing the concerned personnel about the status of the parameters that are needed to monitor and control at the earliest as possible.

4. Hardware Design

We have designed our system in such a way that it provides a complete knowledge of multiple process parameters that are needed to be monitored and controlled. Generally, the sensors output is an analog signal needed to be converted into digital signal via ADC (Analog to digital converter) before interconnecting to the FPGA, which was interfaced with the PC through RF transceiver and monitoring such parameters as well. If the situation is such that a certain parameter has reached to an alarming level, it will be automatically control by the main controller after a certain delay if the control actuation is not provided by that user.

GSM has interfaced with the PC in order to send the information messages to the concerned personnel if the undesired level has reached by a process parameter and at the same time we have incorporated a feature as shown in figure (3) in our proposed system that it will uploading such parameters online continuously after every 5 minutes.

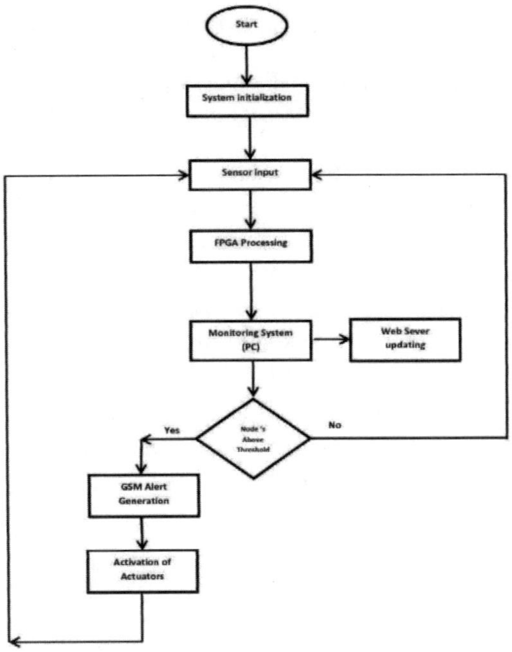

Fig. (3) shown the flowchart of information uploading system

In the mean while if the dramatic change in parameters occurred within this duration, then the message will be delivered to the concerned maintenance team at plant to handle these situations at the earliest while these undesired values are stored in the online data base to keep the record of such events occurred during the plant operations. A block diagram of our proposed system has shown in figure 4.

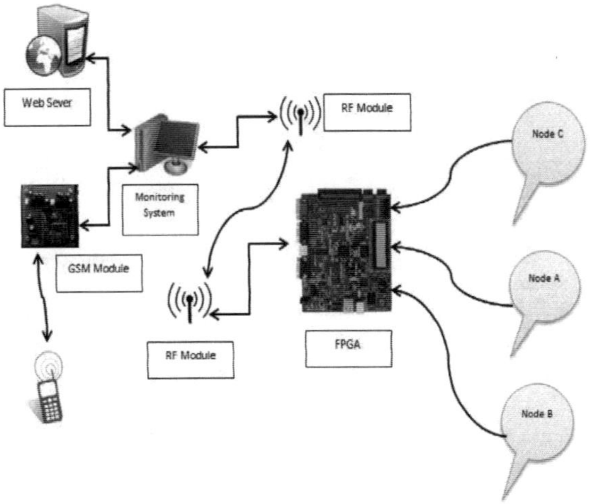

Fig. (4) System Block Diagram

5. Results

We have tested and simulated our proposed system at room temperature. To investigate the performance of our designed system using different sensors configured and adjusted according to the multiple inputs and multiple outputs operational environments. Some tested values from few sensors are shown in table 1 below. Moreover, we have also tested the GSM services incorporated for our system that is for example if the Temperature reaches above the threshold level the incorporated GSM system generated a message to the concerned personnel for information and necessary actions about the situation. In case, if the personnel are outside the plant then they can also control the actuator via message if needed

Table (1): Node A is LM35, Node B is Pyrometer, and Node C is PT-100 showing some operational sensors respectively.

S.NO	Time(min)	Node A (0C)	Node B (0C)	Node C (0C)
1	05	25	611	25
2	10	26	672	27
3	15	27	712	27
4	20	26	733	26
5	25	25	751	25
6	30	30	788	29

In table (1), the node A is showing temperature sensor LM35 and the change in temperature was monitored after every 5 minute and this time duration is adjustable through software alteration as per required conditions. Same is for the rest of the sensors like PT-100 which is node C and Pyrometer which is node B. The graphical outputs of these few sensors are shown in figure (5) below.

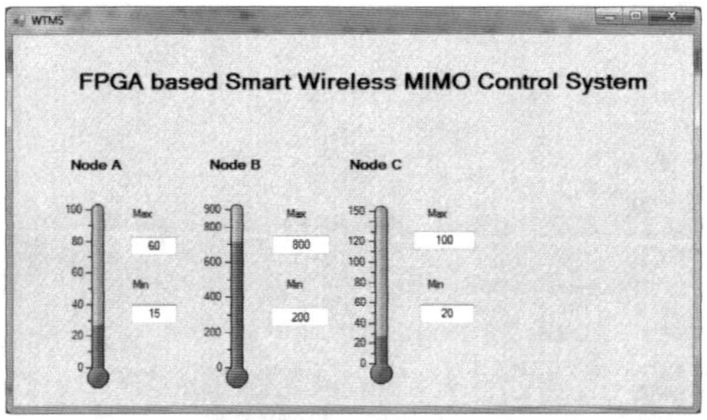

Fig. (5) The graphical outputs of these few sensors

It has been observed during the testing that the proposed system works well according to the industrial requirement. All the parameters are monitored and controlled successfully through this system. This system has an edge over other systems because of its capability for controlling multiple industrial process parameters using FPGA as main controller. As FPGA is capable of parallel processing therefore, this system has fast response time and fast control actuations.

6. Conclusion

We have successfully designed and demonstrated a smart wireless MIMO control system and utilized an FPGA as our main controller capable of parallel processing. Thus, resulted a fast response time for monitoring and control actuations have achieved. Multiple sensors were interfaced with our FPGA for monitoring industrial process parameters such as Temperature, Pressure, and Pyrometers etc. We have also incorporated this feature for controlling such process parameters via messaging through GSM services. Based on all these advantageous features, we believed that this proposed system can be helpful in industrial working environments where conventional monitoring and control is difficult and time consuming and may be hazardous for human presence. Moreover, this system has the flexibility to be monitored through web as well and a data base has also been created to store the occurrence of any suspicious or undesired reading and their remedial action may also be recorded for the future smooth operations.

Reference

[1] DIGILENT, INC. Xilinx XC3S500E FPGA, http://www.digilentinc.com/Products/Detail.cfm?NavPath=2,400,792&Prod=S3EBOARD".

[2] GSM, SIM300 Module, Datasheet, "http://www.google.com/url?sa=t&rct=j&q=&esrc=s&source=web&cd=1&ved=0CC4QFjAA&url=http%3A%2F%2Fprobots.co.in%2FManuals%2FSIM300.pdf&ei=GEVdUYC1HtCP4gTVkoCQDg&usg=AFQjCNFeKRvB8XIXx1kJ4qyzk3CJ3McPBg&bvm=bv.44770516,d.bGE".

[3] Samir Palnitkar, "Verilog HDL A guide to Digital Design and Synthesis," SunSoft Press, 1996.

[4] Prof.S.S.Sarade, Prof A.C. Joshi, Prof Sachin S. Patel, and Prof.A.N. Shinda 2012. Wireless Temperature monitoring system using Wireless Sensor Networks. International Journal of Advanced Research in Electronics and Communication Engineering (IJARECE), Volume 1, Issue 4 October 2012.

[5] Theoplius Wellen and Bhudi Setiawan, "A microcontroller- based room temperature monitoring system," International Journal of Computer Applications (0975 – 8887) Volume 53– No.1, 2012

[6] Hu Bing and Fan Wenyao,"Design of wireless temperature monitoring and control system based on Zigbee technology in communication room," International Conference on Internet Technology and Application (August 2010), 1-3.

[7] Lin Ke, Huang Ting-lei, and Li Lifand, "Design of temperature and humidity monitoring system based on Zigbee technology," Control and Decision Conference (CCDC) June 2009. Chinese 3628-3631, 2009.

[8] Manoranjan Das1 and Banoj Kumar Panda," Prototyping a Wireless Sensor Node using FPGA for Mines Safety Application", ACEEE Int. J. on Electrical and Power Engineering, Vol. 02, No. 02, August 2011.

[9] SUN Yu-jia,WANG Xiao-ming,JIA Fang-xiu, and YU Ji-yan "Design of a Wireless Sensor Network node based on STM32", Proceedings of the 2nd International Conference on Computer Science and Electronics Engineering (ICCSEE 2013).

IOP Conf. Series: Materials Science and Engineering **51** (2013) 012019 doi:10.1088/1757-899X/51/1/012019

Design and Simulation of a Birdcage Coil using CST Studio Suite for Application at 7T

Bernat Palau Tomas, Houmin Li and M R Anjum
Beijing Institute of Technology, Beijing, China.

Abstract. This work describes the study of coils for Magnetic Resonance Imaging (MRI) applications. The principal objective is the design of a birdcage Radio Frequency (RF) coil to use in a 7 Tesla (7T) scanner. Higher strength field generates a better SNR and increased chemical shift effect, improving spectral fat suppression and spectroscopy. Moreover, a better SNR increases the spatial resolution or reduces the imaging time. This research work presented recent developments based on high field 7T design using CST studio. The birdcage coil achieves circular polarization and generates a high homogeneous radio frequency magnetic field under many conditions. Design of a Birdcage coil for a 7T to obtain the images from s mall animals (i.e. mouse). It opens the door to design and construct a Birdcage coil for a 7T to obtain human brain images. Firstly we design a birdcage coil then the results are obtained with simulator CST Wave Studio, creating a 3D model and generating a simulation. Finally the parameters are re adjusted to obtain our desired Larmor frequency 298.2 MHz for a correct operation in 7T. This research work demonstrates the theoretical results from our design and shows the designed antenna behavior.

Keywords: EM simulation, MRI, RF, Antenna

1. Introduction

MRI is a medical non-invasive technique that is used to obtain high quality images of the internal tissues i-e, inside of the human body or animal body. Thus MRI uses the Nuclear Magnetic Resonance property (NMR) to produce images from the nuclei of atoms inside the body. NMR phenomenon has become an indispensable tool in the field of chemistry, biochemistry and structural biological engineering field. At the beginning the imaging technique was called Nuclear Magnetic Resonance Imaging (NMRI) but for the negative connotations associated with the word “nuclear” (nuclear weapons) the name was changed for MRI. In this work the principal objective is to design of a birdcage coil (RF coil) to use in a 7 Tesla scanner. The clinical systems usually operate in the range of 0.2-3T. A more strength field has a better SNR and increased chemical shift effects, improving spectral fat suppression and spectroscopy. Moreover, a better SNR increases the spatial resolution or reduces the imaging time. The birdcage coil

was introduced two decades ago and even it is a popular coil used in MRI because of its ability to achieve circular polarization and generate a high homogeneous radio frequency magnetic field under many conditions. In this research work the design of a Birdcage coil for a 7T to obtain images from small animals, it opens the door to design and construct a Birdcage coil for a 7T to obtain human brain images. The results of the simulations indicate that the design of our RF coil works in the correct Larmor frequency to work in 7T. At the beginning it is necessary to consider some information for specify our antenna, The design of RF coil for use in a 7T scanner (B0).The birdcage design is a Lowpass birdcage coil, The selection is because this type of antenna has less connections and elements than others. This is an important aspect to produce this kind of prototype, because it is less complex to construct and the reduce the cost than the other kind of birdcage models due to less capacitors utilisation.[1,2,3,4,5,6]
The paper is organised as follows, In section 2 describe the simulation result using CST suite, The work ends with Section 3 in which, Conclusion are pointed out.

2. Simulation Result

CST studio suite is an accurate and efficient computational solution for electromagnetic designs. It allows the flexible design and to optimize devices operating in a wide range of frequencies from static to optical frequencies. Moreover, it can analyze the thermal and mechanical effects, as well as, circuit simulation. CST design facilities multi functions and for co-simulation. CST Studio Suite comprises a group of modules, CST Microwave Studio, CST EM studio, CST Particle Studio, CST Cable Studio, CST PCB Studio, CST, CST design studio.

This work is presented the CST Microwave Studio, it is a tool for an accurate 3D simulation of high frequency (HF) devices. It enables a fast and accurate analysis of antennas, filters couplers, planar and multi-layer structure. CST suite offers the time domain solver and the frequency domain solver also it offers solver modules for specific applications. It has filters for the import of specific CAD files and the extraction of SPICE parameters. CST used a mesh system to analyze the 3D models and obtain the simulations.

Table 1 Results obtained Birdcage Builder

Type of leg	Rectangular
Number of legs	8
Type of ER	Tubular
Frequency	298.2MHz
Leg length	16.4
Leg Width	0.5
Coil Radius	9
RF shield radius	10
Capacitance	1pf
Leg self inductance	153.62nH
Er seg self inductance	46.44nH

For simulate the birdcage coil we introduce the 3D model in the CST Wave Studio with the specifications parameter as shown in Table 1 is obtained in the Birdcage Builder and simulation is shown in Fig.1, S-parameters graph obtained with the simulation with CST wave studio shown in Fig.2, Time signal graph obtained with the simulation with CST wave studio shown in fig.3, the readjusted vales in the birdcage coil is shown in table 2, S-parameters graph obtained with this readjusted values simulated using CST wave studio as shown in fig Fig.4. Radiation power in the resonance frequency in the Larmor frequency obtained with the simulation with CST wave studio shown in Fig. 5, S-parameters as a function of frequency graph is obtained with the simulation using CST wave studio as shown in Fig.6 and 3D birdcage coil model with CST wave studio using tetrahedral mesh shown in Fig. 7.

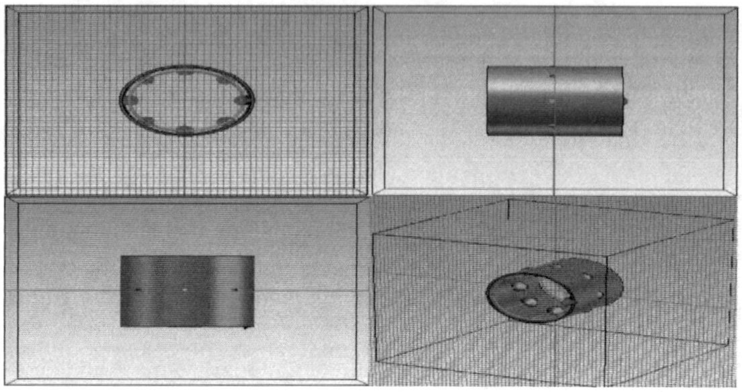

Fig.1 3D birdcage coil model with CST wave studio

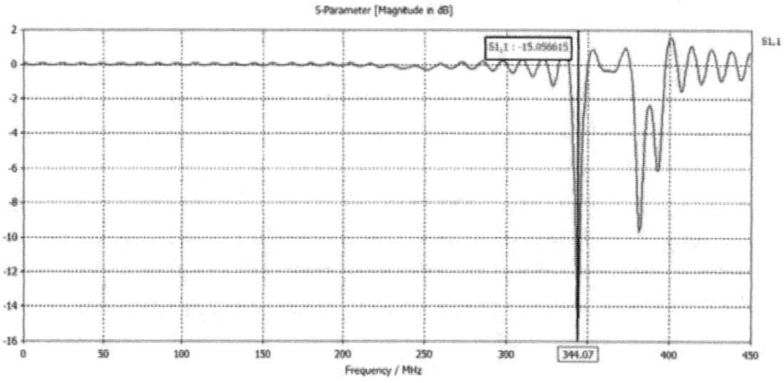

Fig.2 S-parameters graph obtained with the simulation with CST wave studio.

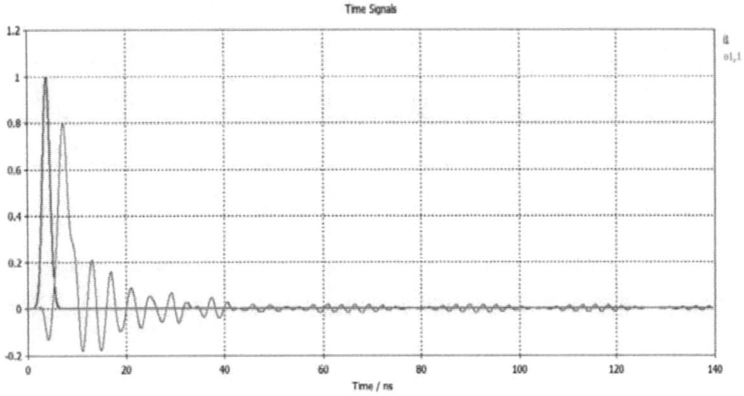

Fig.3 Time signal graph obtained with the simulation with CST wave studio.

Table 2 readjusted values simulated in CST wave studio

Type of leg	Circular	Type of leg	Circular
Number of legs	8	Frequency	298.2 MHz
Type of ER	Tubular	Leg width	0.5
Leg Length	16.4	Coil Radius	9
Leg self induction	153.62nH	Er seg self inductance	46.44nH
RF Shield Radius	10	Capacitance	1PF

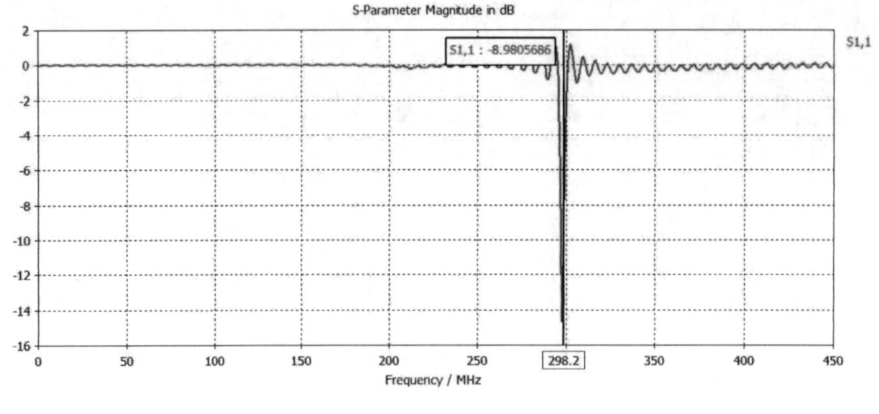

Fig.4 S-parameters graph obtained with the simulation using CST wave studio.

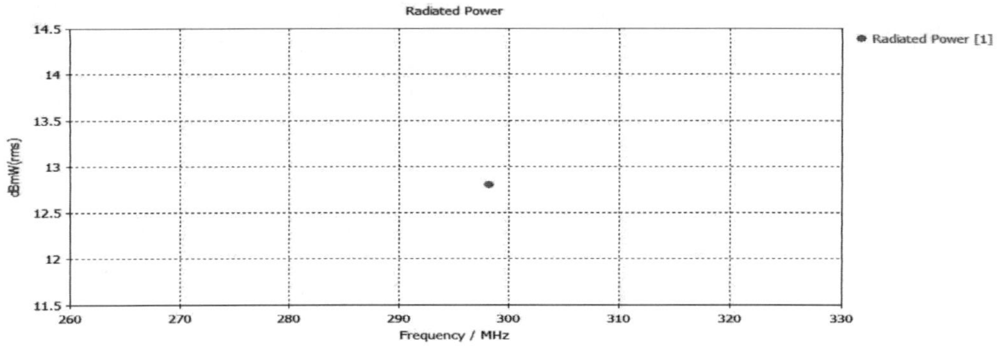

Fig.5 Radiated Power in the Larmor frequency obtained with the simulation with CST wave studio.

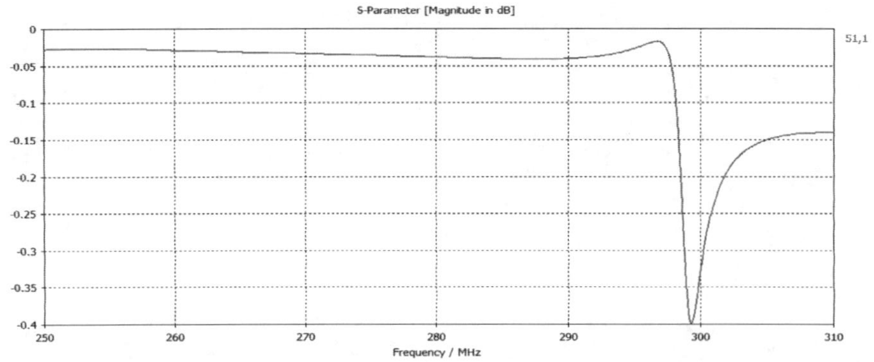

Fig.6 S-parameters graph obtained with the simulation with CST wave studio

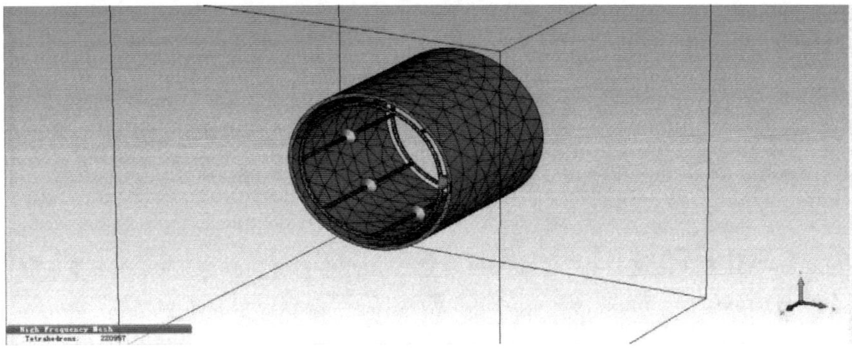

Fig.7 3D birdcage coil model with CST wave studio using tetrahedral mesh

3. Conclusion

This research work obtained the first design and construct a birdcage coil to observe scan image from animals in a 7T, using prototype with the necessary resonance frequency the Larmor frequency for 7T (298.2 MHz). First designed the birdcage using the Birdcage Builder and then simulate it with the CST design studio suite. The first simulations results carried out an error then readapted the design obtained with the Birdcage Builder simulator. Finally using the time solver and the frequency solver we obtained the Larmor frequency. Moreover simulation is obtain a future birdcage coil for generate images from human brain. Finally simulations results have been carried out to verify the effectiveness of the proposed method performed with CST studio suite verify the predictions using the time solver and the frequency solver we obtained the designed Larmor frequency.

References:

[1] Luigi Landini, Vincenzo Positano, Maria Filomena Santarelli "Advanced Magnetic Image in Resonance Processing Imaging" Taylor&Francis Group, Boca Raton, 2005.

ISBN: 9788126522163

[2] J.M.B. Kroot "Analysis of Eddy Currents in a Gradient Coil" Ponsen & Looijen, Wageningen, 2005.

[3] Balk, M.C. "3D Magnetron simulation with CST STUDIO SUITE™," Vacuum Electronics Conference (IVEC), 2011 IEEE International , vol., no., pp.443-444, 21-24 Feb. 2011.

[4] Balk, M.C., Kory, C.L., Dayton, J.A. , "Investigation of a 95GHz helical TWT with CST STUDIO SUITE™," Vacuum Electronics Conference, 2009. IVEC '09. IEEE International , vol., no., pp.505-506, 28-30 April 2009.

[5] Kory, C.L., Dayton, J.A. "Design of 650 GHz helical BWO using CST studio suite," Vacuum Electronics Conference, 2008. IVEC 2008. IEEE International , vol., no., pp.392-393, 22-24 April 2008.

[6] Becker, U. "Recent Developments in 3D Electromagnetic and Charged Particle Simulation" Vacuum Electronics Conference, 2007. IVEC '07. IEEE International , vol., no., pp.1-4, 15-17 May 2007.

[7] CST STUDIO SUITE user's Manual; see also www.cst.com

Sensitivity Analysis of Digital I&C Modules in Protection and Safety Systems

Rahman Khalil Ur, M Zubair, G Heo
College of Engineering, *Department of Nuclear Engineering: Kyung Hee University*, Yongin Si, Gyeonggi Do, 446-701, Republic of Korea

E-mail: gheo@khu.ac.kr

Abstract. This research is performed to examine the sensitivity of digital Instrumentation and Control (I&C) components and modules used in regulating and protection systems architectures of nuclear industry. Fault Tree Analysis (FTA) was performed for four configurations of RPS channel architecture. The channel unavailability has been calculated by using AIMS-PSA, which comes out 4.517E-03, 2.551E-03, 2.246E-03 and 2.7613-04 for architecture configuration I, II, III and IV respectively. It is observed that unavailability decreases by 43.5 % & 50.4% by inserting partial redundancy whereas maximum reduction of 93.9 % in unavailability happens when double redundancy is inserted in architecture. Coincidence module output failure and bi-stable output failures are identified as sensitive failures by Risk Reduction Worth (RRW) and Fussell-Vesely (FV) importance. RRW highlights that risk from coincidence processor output failure can reduced by 48.83 folds and FV indicates that BP output is sensitive by 0.9796 (on a scale of 1).

1. Introduction

Nuclear Industry has lot of application in power generation, water desalination, thermal heat source, testing, well logging, food sterilization and preservation, material research and education. It is therefore essential that all these application must be applied safely and securely in order to keep public, environment and operating staff protected from harmful effects of radiation. Reliability and safety of components and systems used in this industry has become high level objective of prime importance. Commercial nuclear power plants and research reactors can be categorized broadly under the umbrella of nuclear industry. The further discussion in this article will be focused on research reactor.

The primary objectives of research reactors are education, research and analysis, so these are generally designed in the range of several kilo watts to few megawatts. According to IAEA TECDOC-1234, Research reactors with 0.250-2.0 MW power rating or 2.5×10^{11} to $10 \times 10^{11} n/cm^2.s$ flux are termed low power reactor whereas research reactors ranging from 2- 10 MW power rating or 0.1×10^{13} to 10×10^{13} n/cm^2. s are considered as medium to high power research reactors [1]. Instrumentation & Control (I&C) systems used in early designs of research reactors were analogue whereas recently these are either designed using digital ones or being replaced with the digital I&C. The novel kind of issues has been observed with the use of digital I&C, which were not relevant in case of analogue. The salient issues are (1) weakness of digital systems for Common Cause Failure (CCF), (2) threat to cyber security and inter-channel communication, (3) need for highly integrated control room and (4) difficulty to assess the digital I&C reliability [2].

Research objective, in this article, is sensitivity analysis of I&C components and modules in Instrumentation and Control (I&C) architecture. In order to achieve this objective, the preventive maintenance optimization process has been developed to reduce unavailability and different architecture configurations were formulated and their fault trees were developed using AIMS PSA to estimate the unavailability. The failure combination or Minimal Cut Sets (MCS) related to Bi-stable Processor (BP) and Coincidence Processor (CP) are monitored and their risk importance measures are estimated in this study. The current article presents the analysis of four architectures.

2. Literature Review

Safety analysis of nuclear facilities specially research reactors and power plants can be divided into two categories viz. Deterministic Safety Analysis (DSA) and Probabilistic Safety Assessment (PSA). Regulatory bodies and operating organizations in nuclear operating nations have to perform both these analysis. IAEA is playing a key and leading role in these areas specially PSA. IAEA experts evaluated reliability data sources [3] analysed and complied the component reliability data for PSA [4]. United State National Regulatory Commission (USNRC) performed a lot of study in the field of PSA. USNRC performed a reactor safety study jointly with education institutes in 1975 [5] and published a detail PSA procedure guide in 1983 [6]. USNRC consultants are working in the area of digital I&C systems [7] and industry averaged failure data analysis based on initiating events of PSA study [8].

Research institutes and design companies are also focusing on PSA, as it can be used in risk based designing and operation. PSA has ability to identify the importance of I&C systems in nuclear facilities based on risks. Risk significance deriving sensitivity analysis of reactor trip system due to software induced common cause failure was performed by Shahabeddin et al [9]. M. Zubair et al performed quantitative and qualitative analyses of safety parameters by using an online risk monitor system (ORMS), which can check the availability of I&C systems and monitor it during sever accident [10]. Human reliability analysis is an important constituent of PSA and human error probabilities are highly contributing failures during operation and maintenance activities. Poong et al attempted to introduce advanced Man Machine Interface System (MMIS) which will be reduce human error probability [11]. In this article, I&C component based sensitivity analysis was performed first time and it is a novel concept to identify the risk significant components instead of system and addressing them in designing I&C architecture, which will be leading towards risk based designing approach.

3. Analysis technique

Probabilistic Safety Assessment (PSA) is a systematic approach which is used in nuclear, chemical and aviation industry to analyze the system failure in terms of system unavailability and to predict the failure scenarios. In PSA it is also important to identify primary function of each component. In nuclear industry, PSA is used to validate the design of nuclear power plant against set of Postulated Initiating Events (PIEs) [12]. PSA analysis frame work consists of Initiating Event Analysis, Accident Sequence Analysis, and System Analysis etc. Since this article is focused on sensitivity analysis of a system, so Fault Tree Analysis (FTA) technique was opted for system analysis. Fault tree is a good technique for the analysis of complex system, as it yields cut sets showing the failure of components in combinations.

The tool, used to perform Fault Tree Analysis, is AIMS-PSA package. This tool is developed by Korean Atomic Energy Research Institute (KAERI). AIMS software performs the analysis of risk models that are composed of event tree and fault tree. It is used in risk evaluations of nuclear plant's PSAs or risk assessment of chemical plants.

4. System Configuration

Instrumentation and Control (I&C) systems are designed to control, protect and monitor the safe operation of commercial power plants and research reactors in order to maintain safety functions:

a) Controlling the reactivity in the core,

b) Cooling the core to avoid Departure from Nucleate Boiling Ratio (DNBR) violation,

c) Confining the radioactivity

Reactor Control System is responsible for regulating the reactor power level using Control Rods (CRs) within set margin of operation during normal operation. It can perform the reactor control function including reactor start up, changing power levels, and maintaining the power at required level. Reactor Protection System (RPS) is a scram system that initiates the reactor trip to protect the core by inserting shutdown control rods, when operating parameters exceeds the set point or external hazard happens. Monitoring system is responsible to monitor the reactor parameters after and before the design basis accident. The system of interest in this research is reactor protection system and from now on architecture description will be related to RPS.

Architecture of RPS consists of field sensors, Bi-stable Processors (BP), coincidence logic, and circuit breakers to release control rods. Reactor protection system has inter-channel redundancy as well as intra-channel redundancy and this redundancy varies from plant to plant based on design features. Usually RPS consists of four channels in high power reactors with trip logic of 2/4 whereas three redundant and physically separated channels may work 2/3 trip logic for small or low power reactors. Instead of overall reactor protection system, only architecture of a single channel is articulated with four different configurations to see sensitivity effect.

4.1. Architecture Configuration-1

RPS single channel architecture configuration-1 is shown in Figure. 1. Configuration-1 consists of a single bi-stable processor BP_A and single coincidence processor CP_A and circuit breakers to trip with 2/3 logic. This configuration is very basic for a channel and has no inter-channel redundancy. Inter-channel redundancy means redundant modules within a channel whereas intra-channel redundancy is based on number of channels.

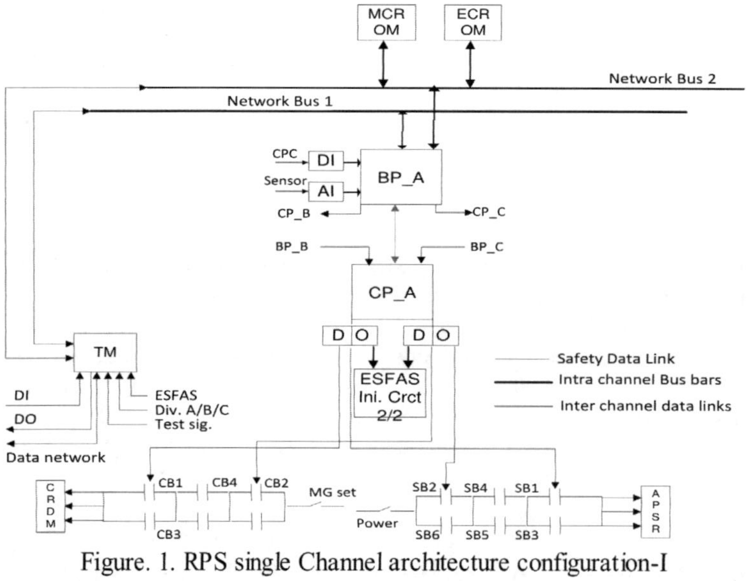

Figure. 1. RPS single Channel architecture configuration-I

4.2. Architecture Configuration-II

In architecture configuration-II, redundancy is added in bi-stable processors to evaluate the impact on single channel. Configuration-II is depicted in Figure.2. RPS channel configuration –II consists of a bi-stable processor BP_A1 & BP_A2, a CP processor CP_A and circuit breakers to trip with 2/3 logic.

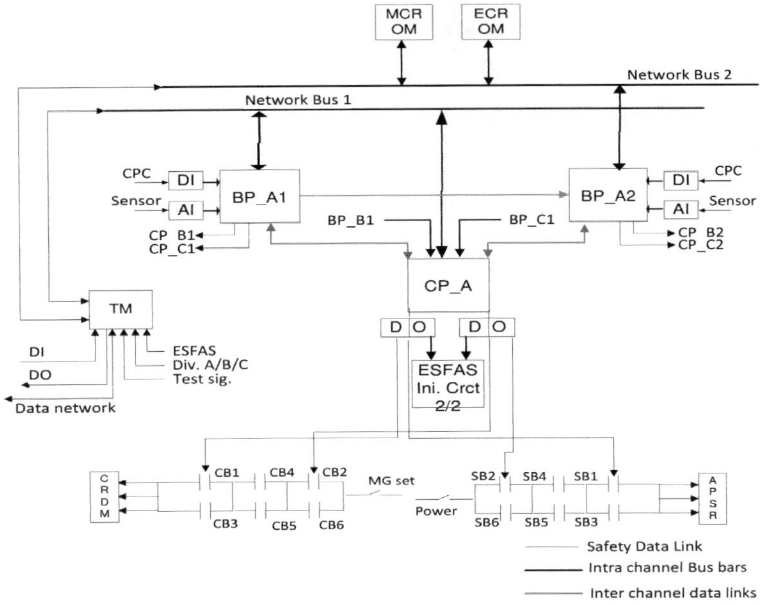

Figure. 2. RPS single Channel architecture configuration-II

4.3. *Architecture Configuration-III*

In order to observe the sensitivity of CP module on single channel failure, coincidence processors is added in the channel, the configuration-III is given in the Figure. 3. RPS channel configuration –II consists of a bi-stable processor BP_A, redundant pair of CP processors CP_A1 & CP-A2 and circuit breakers to trip with 2/3 logic.

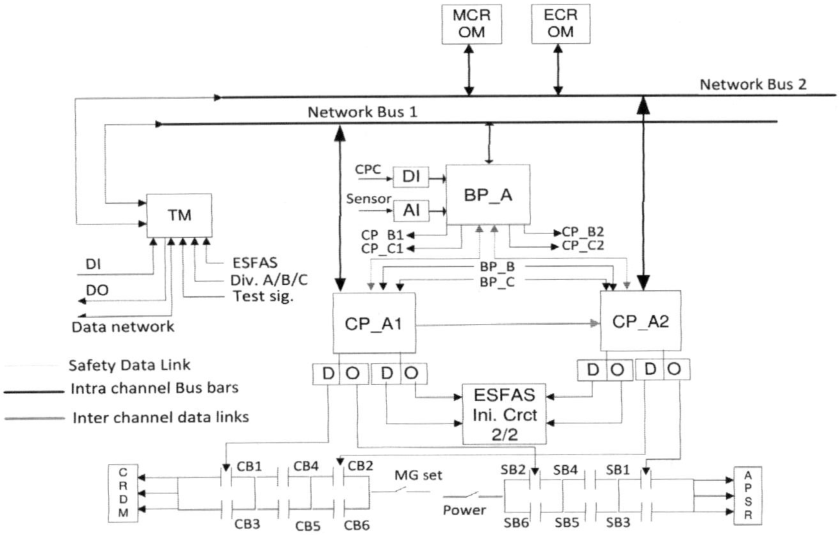

Figure. 3. RPS single Channel architecture configuration-III

4.4. Architecture Configuration-IV

Architecture with inter-channel redundancy of BP &CP modules is shown in Figure. 4. In this configuration, channel architecture consists of redundant bi-stable processors BP_A1 & BP_A2, redundant pair of CP processors CP_A1 & CP-A2 and circuit breakers to trip with 2/3 logic.

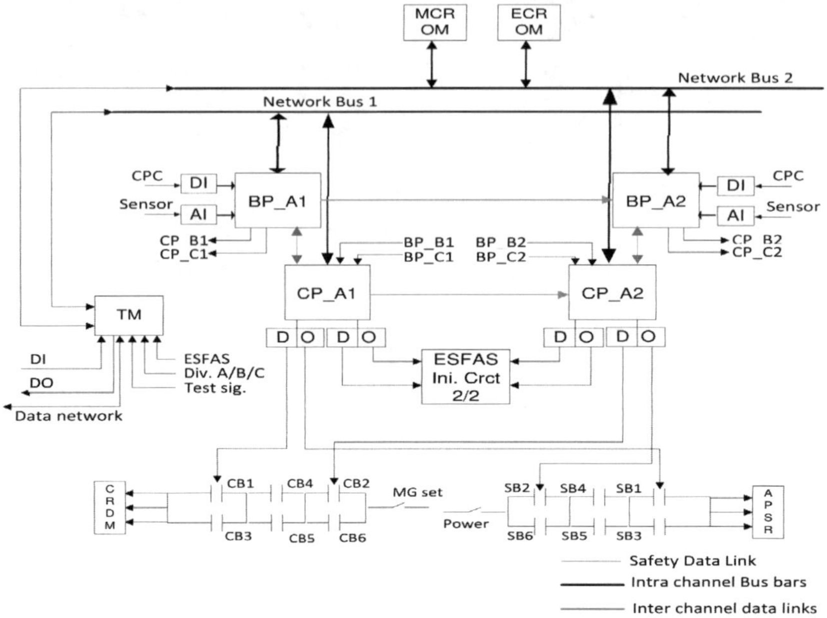

Figure. 4. Channel Architecture Configuration-IV

5. Modelling and failure data

The function of each channel of RPS is to initiate a signal if the input from digital input coming from the core protection calculator (CPC) and analogue sensor input from plant field exceeds the set point parameters. The signal coming from CPC or sensor is compared with set point value in bi-stable processor of each channel and trip signal and/or engineered safety initiation signals are generated. This function failure is modelled as a top gate failure in the fault tree. The interfaces of single channel are with (1) other channels, (2) engineered safety feature actuation system (ESFAS), (3) testing module (TM), (4) control cabinets through network buses and (5) field sensors and digital inputs. The modelling of TM module and ESFAS module is ignored in this analysis.

Success Criteria: The channel is said to be successful, if it generates 3/4 digital output signals from CP. The time window used for quantification of running failures is 372 hours It is based on the surveillance and testing period , which is assumed to be one month and its half (T/2), is taken as time window.

CCF Modelling: Beta factor failure model is for circuit breakers common cause failure and output signal failure from coincidence processor modules. It inserts some conservativeness in the result, which may be changed to MGL or other model in future.

Failure Data: Generic and online available databases are used for this analysis, as shown in table I. The salient reference databases are NUREG/CR-5500 [13], IAEA TECDOC 0930 [14], NUREG/CR-4550 [15], IAEA TECDOC 504 [3] and failure data reports [16].

6. Risk Importance Measures

Risk importance analysis is a sensitivity study which is performed to find the impact of a single failure or combination of failures of basic events on the overall failure of system. Using importance measures,

risk significant basic events can be identified. The sensitivity study differentiates basic events, components into high risk significant and low risk significant components and helps the utilities to focus on their surveillance & maintenance cost management [9].
Common importance measures are of many kinds like Birnbaum sensitivity measure, Improvement Potential, Risk Achievement Worth (RAW), Risk Reduction Worth (RRW), Criticality Importance and Fussell-Vesely (FV) measures. Few of them are discussed in the current article. The risk significance sensitivity criteria are defined in table II by American Society of Mechanical Engineers (ASME).

Table 1. Reference failure data used in FTA

Component	Failure Mode	Failure Probability	Source
Bi-Stable	Fails to Provide Output	7.10E-06	IAEA TECDOC-930
	Module output failure	2.20E-03	NUREG/CR-6579
	Bi-stable hardware failure	3.89E-07	Generic Database
Coincidence Logic Module	Fails to Provide Output	7.10E-06	IAEA TECDOC-930
	Module out failure	2.20E-03	NUREG/CR-6579
Circuit Breaker	Relay fail to open	1.00E-04	NUREG/CR-4639
Network Bus	No update from bus	6.00E-05	EUT Report 93-E-276 [8]
I&C Transmitters	Output Fails during Operation	4.06E-5/h	IAEA TECDOC-930

6.1. Fussell-vesely (FV) Importance

FV importance is a relative measure of failure contribution of a basic event to overall system failure. In simple calculations, it is percentage failure contribution from a basic event. In terms of minimal cut sets, FV is the ratio of unavailability Fi (x) of ith cut set to total unavailability of system, as defined in equation (1).

$$FV_i = \frac{\textit{Sumation} \text{ of cutsets with ith basic event}}{\text{Total system failure/unavailability}}$$

$$FV_i = \frac{F_i(x)}{\sum_{i=1}^{n} F_i(x)} \quad (1)$$

FV is direct indication of the unavailability of the component in the system analysis. Larger the FV importance, the higher component would be sensitive to risk.

Table 2. ASME Sensitivity Criteria for Risk Significance

Sensitivity Parameters		Risk Significance
FV	RAW	
> 0.005	> 2.0	High
< 0.005	< 2.0	Low

6.2. Risk Achievement Worth (RAW)

Risk Achievement Worth is defined as factor by which the unavailability will increase, if the failure of component or basic event is certain. RAW helps to identify the most crucial and critical components or failure modes, which are significant with respect to risk.

$$RAW_i = \frac{F(1)}{\sum_{j=1}^{n} F_j(x)} \tag{2}$$

Where

$\sum_{j=1}^{n} F_j(x)$ = total unavailability of system that is summation of unavailability of all cut sets.

F (1) = total unavailability with basic event probability or cut set unavailability set equal to 1.0.

6.3. Risk Reduction Worth (RRW)

RRW measures the amount by which total risk or unavailability will decrease, if the failure of a basic event or minimal Cut Set (MCS) is zero. RRW also helps to identify the critical components or failure modes; if these are successful system reliability would increase.

The mathematical form of RRW is given by equation (3).

$$RRW_i = \frac{\sum_{j=1}^{n} F_j(x)}{F(0)} \tag{3}$$

Where

$\sum_{j=1}^{n} F_j(x)$ = total unavailability of system that is summation of unavailability of all cut sets.

F (1) = total unavailability with basic event probability or cut set unavailability set equal to 0 (granted success).

7. Results and Discussion

Fault Tree Analysis was performed for four configurations of RPS channel architecture, given in Fig. 1-4. The channel unavailability for each architecture calculated using AIMS-PSA package are elaborated in table III, these are 4.517E-03, 2.551E-03, 2.246E-03 and 2.7613-04 respectively. It can be observed that unavailability decreases by 43.5 % & 50.4% by inserting partial redundancy whereas maximum reduction of 93.9 % happens when double redundancy is added in architecture configuration IV.

Table 3. Unavailability Results of Architecture configurations

case	Channel Unavailability	% Decrease in Unavailability	Configuration
I	4.52E-03	Reference	I BP +1 CP
II	2.55E-03	-43.5	2 BP +1 CP
III	2.24E-03	-50.4	I BP +2 CP
IV	2.76E-04	-93.9	2 BP +2 CP

The top Cut Sets (CS) of each configuration are considered for risk importance analysis. Risk importance analysis in terms of FV, RAW and RRW is calculated in table IV. The salient features of importance analysis are as follow:

a) Importance Analysis of top cut sets of architecture configuration –I shows that both BP and CP module are of equal importance and their output failures are more sensitive to risk. It can be observed from the results depicted in first two rows of table IV.

b) Top minimal cut set from each architecture configuration II &III highlights the output failure from CP and BP module respectively. Sensitivity parameters depict that both failure modes are significant. Failure of CP output is more sensitive.
c) In double redundant architecture, common cause failure of bi-stable processor modules is dominant failure mode and sensitive to risk.

Table 4. Sensitivity Analysis of top cut sets

case	Cut Sets	P(cut set)	FV	RAW	RRW
	RPS-CHA-CFG0-BP-DO-F	2.20E-03	0.4871	221.90	1.95
I	RPS-CHA-CP-CFG0-DO-F	2.20E-03	0.4871	221.90	1.95
	RPS-CHA-CP-CFG0-HD-F	8.52E-05	0.0188	222.37	1.02
II	RPS-CHA-CPA-DO-F	2.20E-03	0.8627	445.26	48.83
	RPS-CHA-BPA12-DO-CCF	2.20E-04	0.0862	446.14	1.11
	RPS-CHA-BP-DO-F	2.20E-03	0.9796	392.29	7.29
III	RPS-CHA-CPA1-AB1AB2-CCF	2.79E-05	0.01242	393.15	1.01
	RPS-CHA-CPA2-CN12-CCF	6.00E-06	0.0026	393.15	1.00
	RPS-CHA-BPA12-DO-CCF	2.20E-04	0.7968	3622.08	4.92
IV	RPS-CHA-CPA1-AB1AB2-CCF	2.79E-05	0.1010	3622.78	1.11
	RPS-CHA-CPA2-CN12-CCF	6.00E-06	0.0217	3622.85	1.02

RRW and FV importance of cut-sets of all architecture configurations is plotted against the cut set numbers, as shown in the Figure. 5. The cut sets corresponding to cut set numbers are given in table V. It can be observed that highest RRW is from cut set # 5 which is of the order of 50, indicating that the risk will reduce by factor of fifty if CP output is reliable. On contrary to RRW, FV importance highlights cut set # 4 (CS-4) as sensitive failure, which related to output failure of BP. Moreover FV indicates that CS-2 to CS-4 has high importance. CS-2, CS-3 & CS-4 are related to BP failure modes while CS-5 and CS-6 are CP failure modes.

Table 5. Sensitivity Analysis of top cut sets

Cut set #	RRW	FV	Cut set	Remarks
CS-1	1.11	0.0862	RPS-CHA-BPA12-DO-CCF	
CS-2	1.95	0.4871	RPS-CHA-CFG0-BP-DO-F	Cut sets/Basic Events related to BP
CS-3	4.92	0.7968	RPS-CHA-BPA12-DO-CCF	
CS-4	7.29	0.9796	RPS-CHA-BP-DO-F	
CS-5	48.83	0.8627	RPS-CHA-CPA-DO-F	
CS-6	1.95	0.4871	RPS-CHA-CP-CFG0-DO-F	
CS-7	1.11	0.1010	RPS-CHA-CP-AB1AB2-CCF	Cut sets/Basic Events related to CP
CS-8	1.02	0.0188	RPS-CHA-CP-CFG0-HD-F	
CS-9	1.02	0.0217	RPS-CHA-CPA2-CN12-CCF	

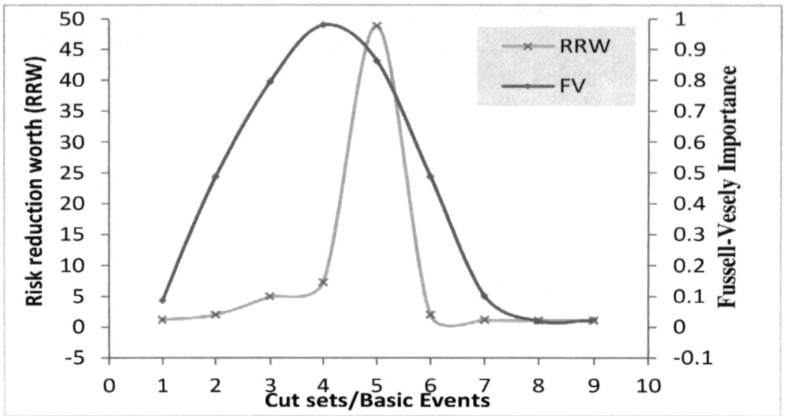

Figure. 5. Risk Reduction Worth and FV importance of cut sets

8. Conclusion

The study has been conducted to analyze risk importance of BP and CP modules and failure modes related to them. It is found that output failure from BP and CP modules is highly significant to risk. RRW highlights that risk from coincidence processor output failure can reduced by 48.83 folds and FV indicates that BP output is sensitive by 0.9796 (on a scale of 1).

In order to cope with this situation, if the redundancy is added to these components, direct failures reduce but common cause failure mode becomes prominent mode of failure, as highlighted by CS-1, CS-3, CS-7 and CS-9 of table V. The study will be continued to analyze unavailability in future, but two additional factor cost and safety category of components would be added to find an optimized architecture for research reactors and remedies to these sensitive failures would be found out.

Acknowledgment

This work was supported by Advanced Research Center for Nuclear Excellence (ARCNEX) program funded by the Ministry of Education, Science and Technology of Republic of Korea (Grant Number: 2011-0031773).

References

[1] International Atomic Energy Agency (IAEA) 2001 *IAEA TECDOC-1234* Vienna
[2] Rahman KU, Heo G, Son H, Kim Y, Park J 2012 *Transactions of the Korean Nuclear Society* 979-80 Korea
[3] International Atomic Energy Agency (IAEA) 1988 IAEA TECDOC-504, Vienna
[4] International Atomic Energy Agency (IAEA) 1988 IAEA TECDOC-478, Vienna
[5] U.S National Regulatory Commission 1983 NUREG/CR-2300 Washington
[6] U.S National Regulatory Commission 1975 WASH-1400, Virginia
[7] U.S National Regulatory Commission 1998 NUREG/CR-6579, Washington
[8] U.S National Regulatory Commission 2007 NUREG/CR-6928, Washington
[9] Kamyab S, Reza M, Nematollahi, Shafiee G 2013 Annals of Nuclear Energy 57 294-303
[10] Zubair M, Ahmed R, Heo G 2013 International Journal of Energy Research ER.3080
[11] POONG HYUN SEONG et al, 2013 Nuclear Engineering and Technology, VOL.45 NO.2, NET.04.2013.700
[12] International Atomic Energy Agency (IAEA) 2010 Specific Safety Guide SSG-3, Vienna

[13] U.S National regulatory Commission 1998 Reliability Study NUREG/CR-5500, Washington
[14] International Atomic Energy Agency (IAEA) 1997 IAEA TECDOC-0930 Vienna
[15] U.S National regulatory Commission 1990 NUREG/CR-4550 Washington
[16] Bollen MHJ 1993 Literature Search for Reliability Data of Components in Electric Distribution Networks, EUT Report 93-E-276 (ISSN 0167-9708), Eindhoven University of Technology, Netherlands

IOP Conf. Series: Materials Science and Engineering **51** (2013) 012021 doi:10.1088/1757-899X/51/1/012021

Longitudinal measurement of chromatic dispersion along an optical fiber transmission system with a new correction factor

Madiha Abbasi[1], Mirza Imran Baig[2] and Muhammad Shafique Shaikh[3]

[1] Dept. of Telecommunication Engineering, Sir Syed University of Engineering and Technology, Karachi, Pakistan (madihaabbasi2@hotmail.com)
[2] Dept. of Telecommunication Engineering, Sir Syed University of Engineering and Technology, Karachi, Pakistan (imran101b@yahoo.com)
[3]Dept. of Electrical and Computer Engineering, King Abdulaziz University, Jeddah, Saudi Arabia (msmuhammad@kau.edu.sa)

Abstract. At present existence OTDR based techniques have become a standard practice for measuring chromatic dispersion distribution along an optical fiber transmission link. A constructive measurement technique has been offered in this paper, in which a four wavelength bidirectional optical time domain reflectometer (OTDR) has been used to compute the chromatic dispersion allocation beside an optical fiber transmission system. To improve the correction factor a novel formulation has been developed, which leads to an enhanced and defined measurement. The investigational outcomes obtained are in good harmony.

1. Introduction

As the fiber optic networks increases rapidly, the responsibilities for monitoring different characteristics of fiber also increases and creating hurdles for telecommunication companies. Various experiment setups for this purpose have been made but actually that are time-taking and costly. For locating breaks and other anomalies, optical time domain reflectometers (OTDR) have been used (1).

Some performance parameters like transmission length and bit rate are badly affected by Chromatic Dispersion (CD). It is produce when different wavelength of light propagates with different velocities in fiber. It raises the pulse broadening effect in transmission systems (1). In dispersion management transmission system, the need for accurate measurement of chromatic dispersion is obvious. By this technique, the chromatic dispersion can be compensated at the wavelength of interest which is helpful in optical networking.

2. Background and theory

2.1. Backscattered Power:

As in (1), the following steps have been carrying out for longitudinal Chromatic Dispersion measurement:

1. Accusation of bi-directional OTDR backscattered power;
2. Estimation of longitudinal mode field diameter (MFD).
3. Estimation of the Waveguide and Material dispersion contribution.

Light which is move all along a fiber with wavelength λ, is backscattered from a generic section z, that backscattered portion is captured by OTDR (1). To know the information about scattered light intensity, the light pulse is taken shorter than the fiber length. The optical power P(λ,z) of backscattered light for a single mode fiber can be expressed as (1), (2) and (3) at time t=0,

$$P(\lambda, z) = P_0 \alpha_S(z) WB(\lambda, z) \frac{c}{N} \exp\left[-2\int_0^z \gamma(x)dx\right] \tag{2.1}$$

where

P_0 = input power
$\alpha_s(z)$ = local scattering coefficient
W = pulse width
$B(\lambda,z)$ = backscatter capture fraction
z = position beside the fiber z = ct/N_2
N = group refractive index
$\gamma(x)$ = local attenuation coefficient

Backscattered power has two parts; one is an exponential decay due to absorption in the glass and scattering. Second signal component is due to the changes in mode field diameter and α_s which is sensitive to local imperfections (1). The resulting trace of OTDR is the summation of these two components. In (4) and (5), separated results of these two OTDR signals have been described.

2.2. MFD Evaluation:

As in (1), S_1(z) is the backscattering trace from the origin and S_2(L-z) is the backscattering trace read from the end of the fiber, which are defined as follows:

$$S_1(\lambda, z) = 10\log[P_1(\lambda, z)] \tag{2.2}$$

$$S_2(\lambda, L-z) = 10\log[P_2(\lambda, L-z)] \tag{2.3}$$

Using equation (2.2) and (2.3), the waveguide imperfection contribution I(λ,z) can be expressed as (1)

$$I(\lambda, z) = \frac{S_1(\lambda, z) + S_2(\lambda, L-z)}{2} \tag{2.4}$$

It is defined in (3) and (6), the mode field diameter 2W(λ,z) has significance as compare to the variation in the local scattering coefficient α_s. Thus, the normalized imperfection contribution $I_n(\lambda,z)$ at z = z_o is as follow

$$I_n(\lambda, z) \equiv I(\lambda, z) - I(\lambda, z_o) = 20\log\left[\frac{W(\lambda, z_o)}{W(\lambda, z)}\right] \tag{2.5}$$

As in (1) and (6), when the mode field diameter 2W(λ,z_o) at z = z_o is given, the mode field diameter distribution 2W(λ,z) can be obtained

$$W(\lambda, z) = W(\lambda, z_o)10^{-\left\{\frac{I_n(\lambda,z)}{20}\right\}} \tag{2.6}$$

2.3. Evaluation of MFD with Correction Factor:

In the MFD Evaluation, the fluctuation in scattering co-efficient $\alpha_s(z)$ across the fiber is assumed negligible in comparison with fluctuation in MFD, from equation (2.6) longitudinal MFD can be obtained. But it is important to take in consideration the length dependency of the fiber when the scattering coefficient differs greatly along a fiber. By considering the change in refractive index and scattering coefficient, the imperfection contribution I(λ,z) expressed in (1), (3) and the normalized imperfection contribution $I_n(\lambda,z)$ at $z = z_o$ can be rewritten as

$$I_n(\lambda, z) = 10\log\left[\frac{\alpha_S(z)\, n^2(z_0)}{\alpha_S(z_0)\, n^2(z)}\right] + 20\log\left[\frac{W(\lambda, z_0)}{W(\lambda, z)}\right] \tag{2.7}$$

The first expression on the right in above equation is considered as correction factor *K*.
It is defined in (1) that by using the relative refractive index Δ and the Rayleigh scattering coefficient *R* for GeO_2 doped core fiber, the correction factor K is rewritten as

$$K = 10\log\left[\left\{\frac{1+0.62\Delta(z)}{1+0.62(z_o)}\right\}\left\{\frac{50-\Delta(z)}{50-\Delta(z_o)}\right\}\right] \tag{2.8}$$

Thus, the distribution of mode field diameter 2W(λ,z) can be acquired as (1)

$$W(\lambda, z) = W(\lambda, z_0)10^{-\left\{\frac{I_n(\lambda,z)-K}{20}\right\}} \tag{2.9}$$

2.4. New Correction Factor K:

The group refractive index of fiber under testing conditions also influenced the received OTDR backscattered power. Therefore by taking in account, it can be figured out from (2.1). The group refractive index expressed as in (7)

$$N = n - \lambda\frac{dn}{d\lambda} \tag{2.10}$$

Where,
n = fiber refractive index
λ = operating wavelength

In (1), an assumption has been made by ignoring the second term in the equation (2.10). By taking into account this neglected factor, a new correction factor has been formulated. Therefore, equation (2.7) can be amended as

$$I_n(\lambda, z) = 10\log\left[\frac{\alpha_s(z)n^2(z_o)(2n(z_o)-1)}{\alpha_s(z_o)n^2(z)(2n(z)-1)}\right] + 20\log\left[\frac{W(\lambda, z_o)}{W(\lambda, z)}\right] \tag{2.11}$$

and the correction factor K has to be changed as

$$K = 10\log\left[\frac{\alpha_s(z)n^2(z_o)(2n(z_o)-1)}{\alpha_s(z_o)n^2(z)(2n(z)-1)}\right] \tag{2.12}$$

Consequently, we get a new correction factor K, which has an effect on the measurement of MFD and chromatic dispersion (CD).

2.5. *Evaluation of Chromatic Dispersion (CD)*:

The CD σ is a summation of the material dispersion σ_m and waveguide dispersion σ_w. σ_m and σ_w calculations and estimations are well defined in (2) and (8) and (9).

$$\sigma_W = \frac{\lambda}{2\pi^2 cn}\frac{d}{d\lambda}\left(\frac{\lambda}{W^2}\right) \tag{2.13}$$

The experiential relationship between the MFD $2W$ and the normalized frequency V has been presented in (10). By substituting the equation of wavelength dependency of mode field diameter $2W$ expressed in (1), in equation (2.13) then the waveguide dispersion σ_w taken from

$$\sigma_w(\lambda, z) = \frac{\lambda}{2\pi^2 cnW^2(\lambda, z)} \times \left\{1 - \frac{2\lambda}{W(\lambda, z)}\left(\frac{3}{2}g_1(z)\lambda^{0.5} + 6g_2(z)\lambda^5\right)\right\} \tag{2.14}$$

3. **Results**

3.1. *Experimental Setup:*

With the OTDR the backscattered signal power is measure by using four wavelengths of 1.31, 1.45, 1.55 and 1.625 μm. The transmission link composed of 2km single mode fiber (SMF), 10 km of SMF, 10 km of SMF and 20 km long dispersion compensation fiber (DCF).

3.2. *Measurement of Mode Field Diameter:*

Figure.1 shows the results for the MFD distribution. The difference between prior and our intended new correction factor K can be observed. Our new results are shown with the red line which is superimposed on previous result which is indicated by blue line. A significant enhancement in the MFD measurements is clear.

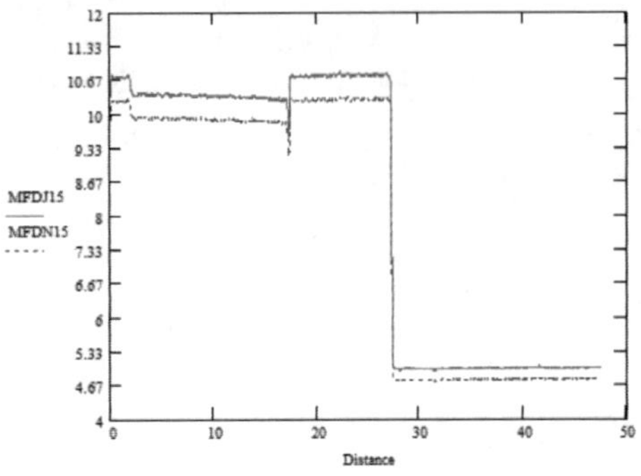

Figure 1. Mode Field Diameter measurement at 1550nm. Blue line represents measurements taken by using previous K and red line shows measurements taken by using new K.

3.3. *Measurement of Chromatic Dispersion*:

The results of longitudinal CD measurement are shown in figure 2. In this paper the both previous and recent result of proposed correction factor K are presented. Result of new correction factor is shown with red color line where as result of old correction factor is represented by blue color line.

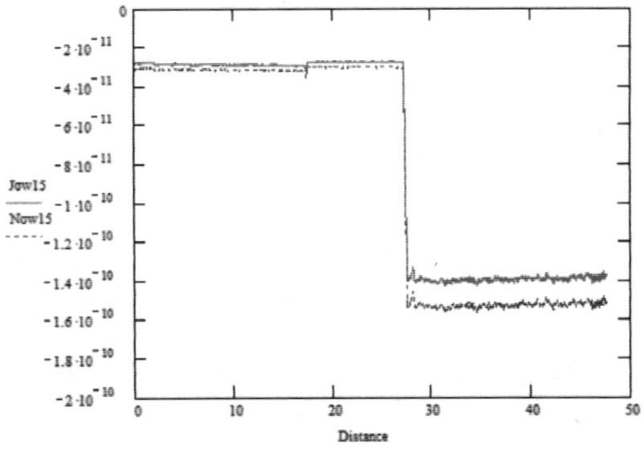

Figure 2. Longitudinal CD measurement at 1550nm. Blue line shows result with previous K and red line shows the result with new K.

4. **Conclusion**

With our projected new correction factor K, the chromatic dispersion measurement along optical fiber transmission system become clearer. As compared with the previous technique with old correction factor, this constructive OTDR based technique with new correction factor gives improved results. The chromatic dispersion measurements are now turning out to be more accurate and precise.

5. **Acknowledgments**

The authors would like to thank Sir syed university of Engineering and Technology for the support and encouragement. Special gratitude to Dr. Mirza Imran baig who supervise and bless with his deep knowledge of fiber optics.

6. **References**

1. "Distributed measurement of chromatic dispersion along an optical fiber transmission system with improved correction factor**", M. Imran Baig, T. Miyauchi and M. Imai, 11, November 2006, Vol. 45.**
2. *"Distributed measurement of chromatic dispersion along an optical fiber transmission system",.* **M. Imai, M. I. Baig and T. Miyauchi,.** 2005. Proc. IEEE'2005/LEOS workshop on fiber and optical passive components. pp. 146-151.

3. *" Chromatic dispersion distribution measurement along a single mode optical fiber",.* **K.Nakajima, M.Ohashi and M.Tateda,.** 7, July 2007, J. of Lightwave Technology,, Vol. 15, pp. 1095-1101.

4. *"Interpretation of SM fiber OTDR signatures", Optical Testing and Metrology.* **M.S. O`Sullivan and R.S. Lowe.** 1986. Proc. SPIE`86. Vol. 661, pp. 171-176.

5. *"Backscattering measurements in optical fibers: Separation of power decay from imperfection contribution".* **P. Di. Vita and U. Rossi.** July 1979, Electron. Lett., Vol. 15, pp. 467-469.

6. *"Novel technique for measuring longitudinal chromatic dispersion in single-mode fibers",.* **M. Ohashi and M. Tateda.** 5, March 1993, Electron. Lett., Vol. 29, pp. 426-427.

7. **John M. Senior.** *"Optical fiber communications principles and practice",.* s.l. : Prentice-Hall, 1985.

8. *"Spatially resolved chromatic dispersion measurement by a bidirectional OTDR technique".* **A. Rossaro, M. Schiano, T. Tambosso and Davide D`Alessandro,.** 3, May/June 2001, IEEE J. on select. topics in quantum electronics, Vol. 7, pp. 475-483.

9. *Material dispersion in lightguide glasses.* **J.W. Fleming.** 11, May 1987, Electron. Lett., Vol. 14.

10. *"Loss analysis of single mode fiber splice".* **D. Marcuse.** 1977, Bell System Technical Journal, Vol. 56, pp. 703-718.

IOP Conf. Series: Materials Science and Engineering **51** (2013) 012022 doi:10.1088/1757-899X/51/1/012022

Model Identification and Validation for a Heating System using MATLAB System Identification Toolbox

Muhammad Junaid Rabbani[1], Kashan Hussain[2], Asim-ur-Rehman khan[3], Abdullah Ali[4]
Electrical Engineering, National University of Computer & Emerging Sciences-FAST Karachi, Pakistan
Email: junaid.rabbani@nu.edu.pk[1], kashan.husain@nu.edu.pk[2], asim.rehman@nu.edu.pk[3], abdullah.ali@nu.edu.pk[4]

Abstract. This paper proposed a systematic approach to select a mathematical model for an industrial heating system by adopting system identification techniques with the aim of fulfilling the design requirement for the controller. The model identification process will begin by collecting real measurement data samples with the aid of MATLAB system identification toolbox. The criteria for selecting the model that could validate model output with actual data will based upon: parametric identification technique, picking the best model structure with low order among ARX, ARMAX and BJ, and then applying model estimation and validation tests. Simulated results have shown that the BJ model has been best in providing good estimation and validation based upon performance criteria such as: final prediction error, loss function, best percentage of model fit, and co-relation analysis of residual for output.

Index Terms: System Identification, ARX Model, ARMAX Model, BJ Model, Validation.

1. Introduction

The idea of this research work initiates from an essential requirement of designing a control system for an industrial heating system used earlier by G. Dullerud and R. Smith [1] and also presented by [2].The heating system as shown in Fig. 1, is a simple Single Input Single Output System (SISO), the input voltage (U) drives a 300 Watt halogen lamp, suspended 2 inches above the metal plate. The output (Y) is a thermocouple measurement, attached underneath the plate to measure heating temperature. While, first step to design a controller required: mathematical model that describe physical properties of a heating system that need to be controlled [3]-[4]-[5]. This paper proposes system identification techniques to determine an efficient mathematical model for a heating system that could validate the real measurement data.

System identification is the class of control system engineering that determines physical characteristics of a plant and represent them in the form of mathematical expression by using real time measurement or experimental data [2]-[6]. After collecting an experimental data, most essential stages of model identification process can be summarized as follows, (1) select between parametric or non parametric modelling technique, (2) choosing the model structure, and finally (3) model estimation and validation criteria [7]. All these stages have a significant impact on model complexity, final prediction error, loss function, and percentage of model fitness.

System identification can be categorized based on their technique such as parametric model identification and non parametric model identification. Non parametric identification technique in comparison with parametric identification technique is relatively simple but less accurate. Whereas, parametric model identification provides complete model description that truly describes the physical dynamics of plant. The model structures for parametric identification are Auto-Regressive Exogenous (ARX), Auto-Regressive Moving Average Exogenous Model (ARMAX), Box-Jenkins (BJ) and State Space Model that are discussed in [8]-[9].

Furthermore, in system identification, model estimation of plant is carried out using measurement data. The objective of model estimation is to minimize the error function between measured and predicted output. Various model estimation approaches can be adopted in parametric estimation such as; prediction error method, least square, instrumental variable and maximum likelihood [9]. Whereas, model validation is vital to find the model credibility whether, model is capable to produce measured data or not. It can be achieved by examine best percentage of model fitting and correlation analysis of residual function between input and output [7].

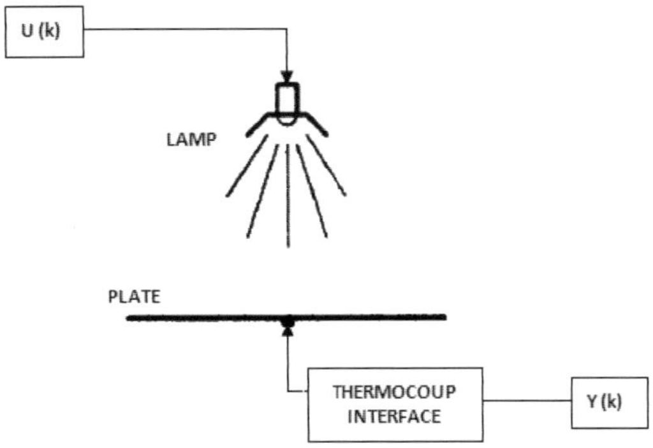

Fig. 1. Heating system block diagram

2. METHODOLOGY

MATLAB System Identification tool box also presented by [10]-[11]-[12] has been used to determine mathematical model for a heating system, which provide ease and computer based simulation with multiple iteration facility to pick the best model that prove maximum model estimation and validation in comparison with real system. This paper used real measurement data, taken directly from the heating system for model identification discussed in [13]. The integral input/output samples data taken at desired sample interval of a heating system are as follow:

U = input voltage (Volt)
Y = Temperature (deg. C)
Sampling interval: 2 seconds
Total number of samples: 801

First load heating system measurement data such as input voltage, sampling time and output temperature in MATLAB and then initialize input signal as (u) vector and output signal as (y) vectors in MATLAB work space. Now open the system identification tool box ('ident') and import input and output data array by setting its starting point as zero and sampling interval to 2.0. Fig. 2 shows the time domain representation of observed data; both output temperature and input voltage is plotted with respect to sampling time period. It can be observed that the relation between input voltage and output temperature is linear. On the other hand, Fig. 2 represents time domain representation of both functions, after subtracting the mean value from the measurement data to remove offset. The reason of subtracting the mean value from each signal is : to build a linear model that would describe deviated response from the physical equilibrium.

Finally the entire measurement data was divided in two parts, keeping the first one reserved for estimation and other for validation. Two-third data from (1 - 533 samples) is selected for estimation and one-third data from (534 - 801) has been reserved for validation. Fig. 3 shows the estimation and validation data set as red and qua line colors respectively. The estimation data are used as an input, and these data sets are input to the model estimator to predict the model as discussed in [2], [3].

This paper compared ARX, ARMAX and BJ model structures to find the best model for a heating system by adopting systematic approach based upon these stages: (i) choosing parametric model identification technique, (ii) picking the best model structure among ARX, ARMAX and BJ model with low order, (iii) apply model estimation and validation techniques by evaluating model performance based upon these performance criteria: (1) minimizing FPE and LF, (2) choose model structure which provide highest percentage of model fit, (3) auto-correlation and cross-correlation analysis of residual for output should be inside the confident region.

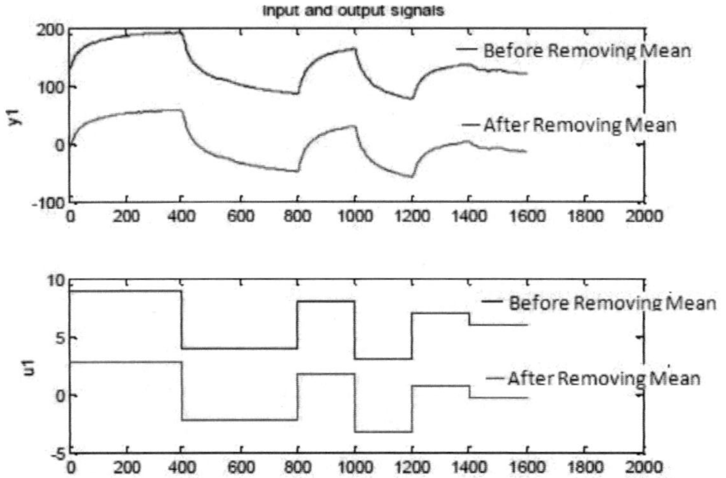

Fig. 2. Heating system real measurement data before and after removing mean

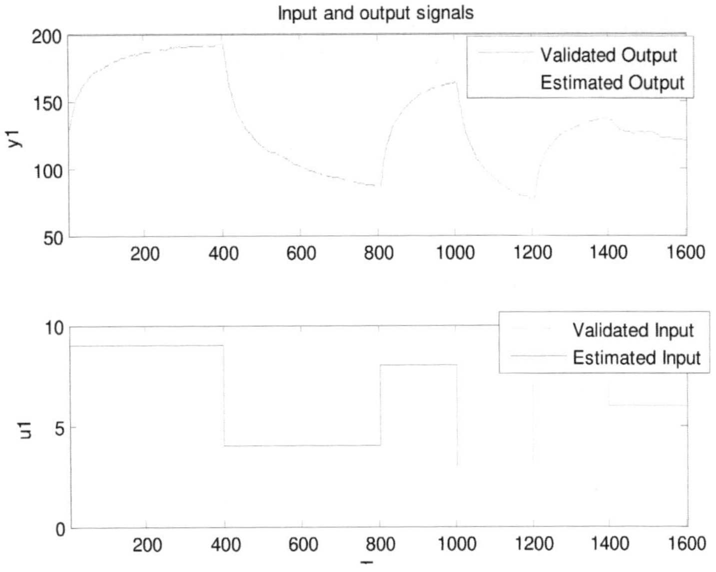

Fig. 3. Estimation & validation data set for heating system

3. MODELING OVERVIEW

A. Auto Regressive Model (ARX)

The ARX model is the simplest model describing input to output path as a linear difference equation. The model output is the sum of input and white noise, which enter directly into the system considered as a measurement noise. The ARX model provides maximum model estimation and validation result while model order id high [3]. The ARX model is defined as:

$$A(q)*y(t) = B(q)*u(t)+e(t) \quad (1)$$

Where A and B polynomials are output and input respectively, And e(t) is the white noise.

$$A(q) = 1+a1q^{-1}+\ldots\ldots+a_{na}q^{-na} \quad (2)$$
$$B(q) = b_1q^{-1}+\ldots..+b_{nb}q^{-nb} \quad (3)$$
$$y(t)+a_1y(t-1)+\ldots..+a_{na}y(t-n_a)=b_1u(t-1)+\ldots..+b_{nb}u(t-n_b)+e(t) \quad (4)$$

B. Auto Regressive and Moving Average Model (ARMAX

The ARMAX model has an advantage over ARX model, because ARX model was not good in describing the property of dynamics disturbance. While ARMAX model includes flexibility to model disturbance separately by adding numerator polynomial C (z − 1). ARMAX model can be useful to handle disturbance, enter at the input [3]. This gives the model:

$$A(q)*y(t)=B(q)*u(t)+C(q)*e(t) \quad (5)$$

Where, A and B polynomials will remain same as in ARX model, C polynomials can be written as

$$C(q)=1+c_1q^{-1}+\ldots..+c_{nc}q^{-nc} \quad (6)$$

C. Box-Jenkins (BJ

Box-Jenkins (BJ) model structure provides further development of output error, by filtering noise through C(q)/D(q) transfer function. The BJ model is useful when noise enter late in the process [3], such as measurement noise.

$$y(t)=\frac{B(q)}{F(q)}u(t)+\frac{C(q)}{D(q)}e(t) \tag{7}$$

4. RESULTS

A. ARX Model Estimation and Validation Results

Firstly, to compute ARX model with suitable order and delay, Fig. 4 shows the best fits for ARX model as a function of: number of parameters used for various combination of A(q) and B(q) polynomials. Fig. 5 shows the percentage of best fit between measured and estimated model output for different A(q) and B(q) polynomials orders. The best estimation for the model is achieved; when the order of A(q) and B(q) polynomials is selected as eight and ten respectively that's give 69.64 % of the best fit. Furthermore, model validation test is done by investigating residual analysis. Figure 6, shows model credibility under the performance of residual analysis. Residual analysis has been done to quantify the error between the predicted output and measured output based on estimated model from the validation data set. Residual analysis consists of both whiteness and independence test to check whether the residuals are uncorrelated and independent from past inputs that prove good model validation. The quality of the estimated model can be judged if the autocorrelation function resides inside the confidence interval comprises of two straight lines. Figure 6, shows the auto-correlation and cross-correlation analysis of residual for ARX model at various orders; it can be observed that residual is inside the confident region (.18 to -.18) for all ARX model order combination. The estimated model equation of the ARX along with the A(q) and B(q) polynomials are as follow,

$$A(q).\, y(t) = B(q).\, u(t)+ e(t) \tag{8}$$

$$A(q) = 1- 1.22\, q^{-1} + 0.1457\, q^{-2} + 0.0451 q^{-3} + 0.18819 q^{-4} + 0.1366\, q^{-5} + 0.08447\, ^{q-6} + 0.0149\, q^{-7} + 0.05336 q^{-8} \tag{9}$$

$$B(q) = 0.2454 q^{-2} + 0.2631\, q^{-3} + 0.1645\, q^{-4} - 0.06359 q^{-5} - 0.1046\, q^{-6} - 0.1062\, q^{-7} - 0.0464 q^{-8} - 0.1641 q^{-9} - 0.09716\, q^{-10} \tag{10}$$

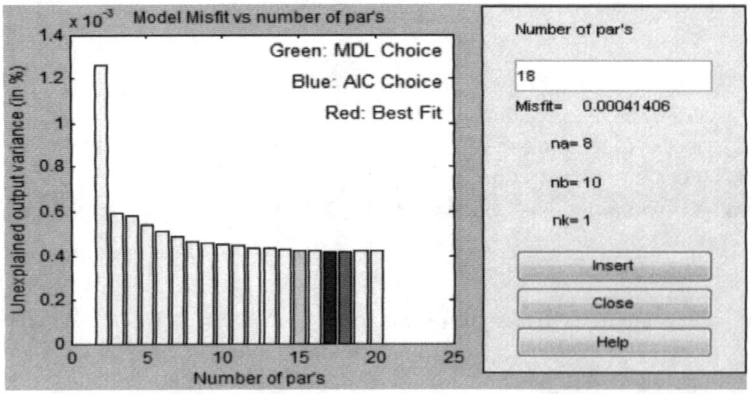

Fig. 4. ARX models best fit for its validation data

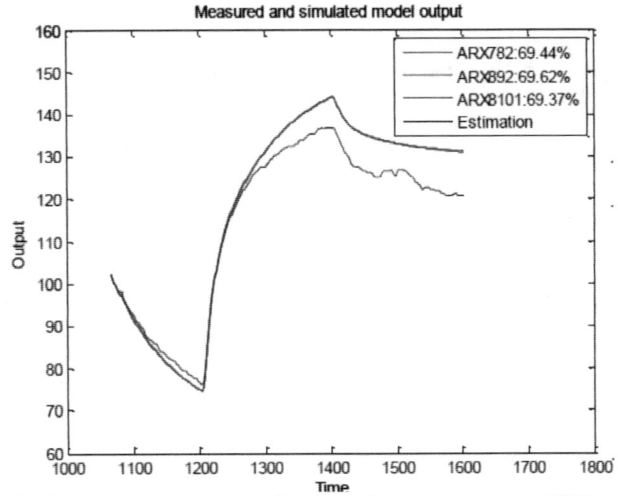

Fig. 5. Best fit % between measured and estimated output at various ARX model order

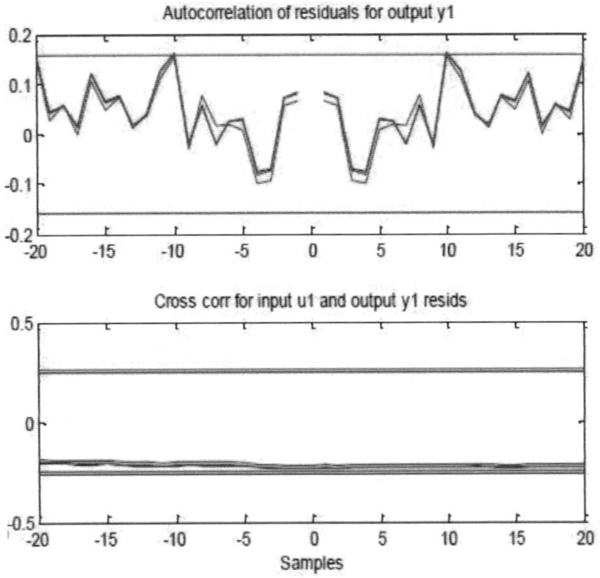

Fig. 6. Co-relation analysis of ARX model for model validation

B. ARMAX Model Estimation and Validation Results

Similarly, ARMAX model has been design to identify the best possible model. Unlike ARX model here, it is also important to find C(q) polynomial along with A(q) and B(q) polynomials to structure noise model as well. Fig. 7, shows best fit between measured and estimated output at different A(q), B(q) and C(q) polynomials orders. Best result is achieved, when order is selected 2, 2, 2 for A(q), B(q) and C(q) polynomials respectively, it provides best possible model fit up to 75% in comparison with other ARMAX models at low order. Figure 8, illustrates the auto-correlation and cross-correlation analysis of residual for ARMAX model at various orders; it can be observed that residual is within the confident range (0.2 to -0.2) for those ARMAX models that fit up to 75 %, ensuring that residuals are not correlated and independent from past input for the desired polynomial order.

While residual lie outside the confident range for ARMAX model estimating up to 65 % model best fit. The estimated mathematical model of heating system using ARMAX model can be written as;

$$A(q)y(t)=B(q)u(t)+C(q)e(t) \quad (11)$$

Where, A, B and C polynomials are as follow,

$$A(q)=1-1.913q^{-1}+0.9141q^{-2} \quad (12)$$
$$B(q)=0.8145q^{-3}+0.7895q^{-4} \quad (13)$$
$$C(q)=1-0.7041q^{-1}+0.009244q^{-2} \quad (14)$$
$$y9t)-1.913y(t-1)+0.9y(t-2)=0.81u(t-3)+0.79u(t-4)+e(t)-0.7e(t-1)+0.0092e(t-2) \quad (15)$$

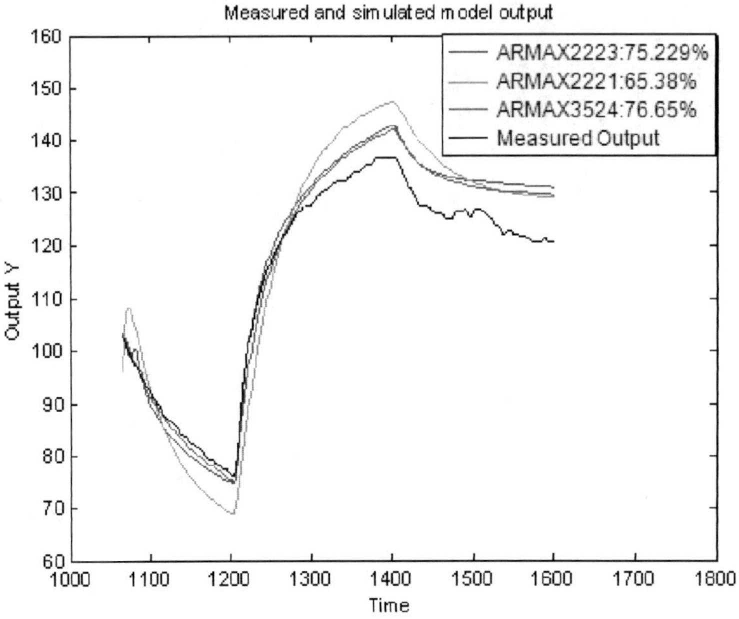

Fig. 7. Best fit % between measured and estimated output at different ARMAX model order

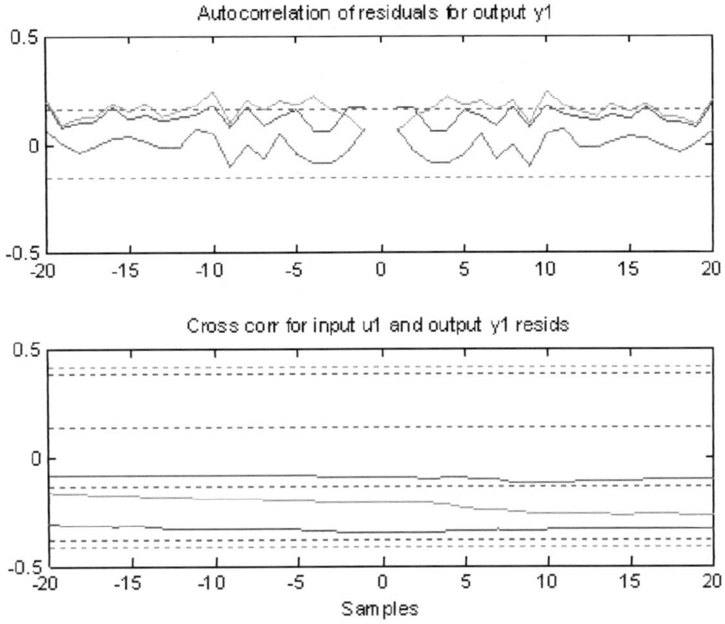

Fig. 8. Co-relation analysis of ARMAX model for model validation

C. BJ Model Estimation and Validation Results

Finally, BJ model has been designed for heating system, Fig. 9, illustrates best fit % between measured and estimated model output. It is noticeable here; BJ model provides maximum model best fit result among all model structures. The efficiency of model is 91.03 %, achieved after choosing B(q), C(q), F(q) and D(q) polynomials order up to 7, 5, 8, 9 respectively. Figure 9, illustrate the auto-correlation and cross-correlation analysis of residual for BJ model structure; it can be observed that again residual is inside the confident region (0.2 to -0.2), with tolerance range ± 1%. It further passes both witness and independence test by indicating: residuals are not correlated and independent from past input that proves the model's capability of validating actual data.

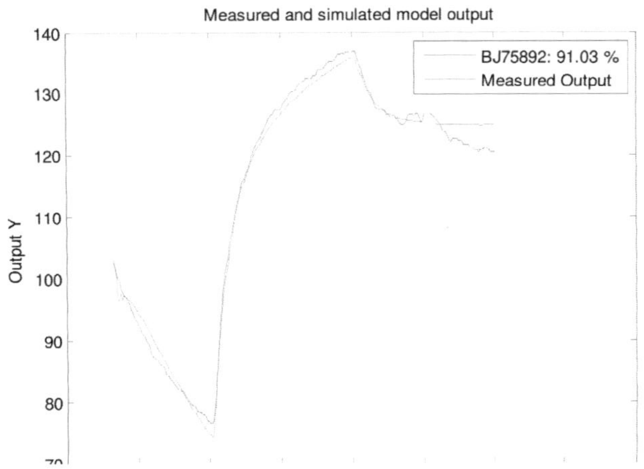

Fig. 9. Best fit % between measured and estimated output of BJ model

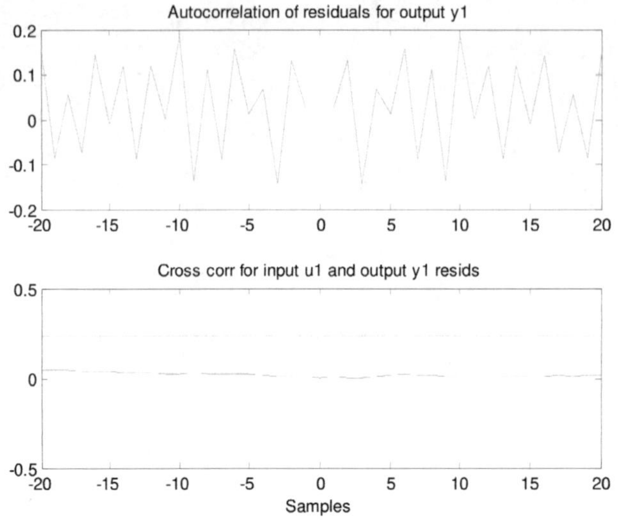

Fig. 10. Co-relation analysis of BJ model for model validation

Figure 11 and Table 1; summarize model estimation and validation results among ARX, ARMA and BJ model structures. It is required to choose best mathematical model for a heating system by analyzing and judging all performance parameters. It can be observed that ARX and ARMAX models provide best fit up to 70 % and 75 % respectively with low FPE and LF. Whereas, BJ model has a tendency to validate model fit up to 91 % as well as provides low FPE and LF. It has also been noticed that all model structures passed validation test under residual analysis.

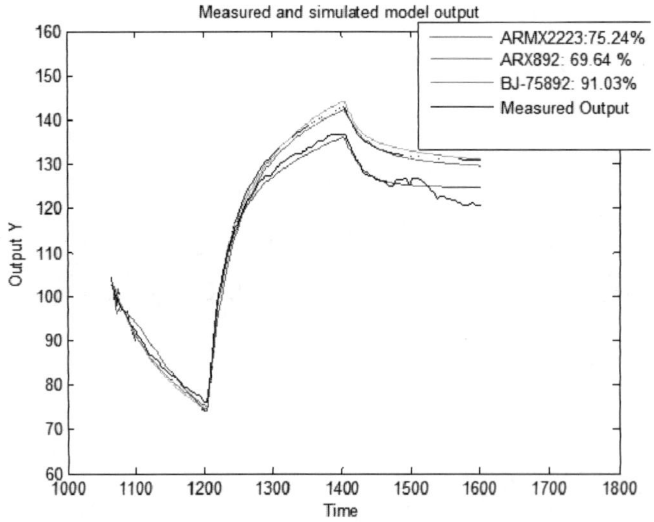

Fig. 11. Best fit % between measured and estimated output of ARX, ARMAX and BJ models

Table. I Model estimation and validation result

Model	LF	FPE	Best fit
			%
ARX (8,9,2)	0.0538576	0.056889	69.64
ARMAX(2,2,2,3)	0.0579158	0.059544	75.24
BJ (75892)	0.054687	0.5243	91.03

5. CONCLUSION

This paper has focused the system identification techniques to select best mathematical model that would be suitable to translate physical behavior of a heating system. Among ARX ARMAX and BJ modeling structure, it seems reasonable to pick BJ model as a final choice; because it provides maximum model estimation and validation up to 91 %, in comparison with ARX and ARMAX models that give 65% and 75% respectively. It also has been observed that all model structures have less FPE and LF as well, passing model validation test under residual analysis. On the other hand, BJ model structure increases model complexity due to increase in system order in contrast with ARMAX model. So, there is a tradeoff between system order and best percentage of model fit.

6. REFERENCES

[1] G. Dullerud and R. Smith, "The experimental validation of robust control models for a heat experiment: A linear matrix inequality approach," in Decision and Control, 1995., Proceedings of the 34th IEEE Conference on, 1995, pp. 3486-3491.

[2] T. Mulyana, M. Than, M. Nor and N. A. Mustapha, "Identification of heat exchanger qad model BDT 921 based on hammerstein-wiener model," in International Seminar on the Application of Science & Mathematics 2011.

[3] L. Ljung, System identification - Theory for the User, Prentice-Hall, USA, 1999.

[4] T. Söderström and P. Stoica, System Identification. Prentice-Hall, Inc., 1988.

[5] R. Pintelon and J. Schoukens, System Identification: A Frequency Domain Approach. Wiley-IEEE Press, 2004.

[6] M. Gontikaki and A. van Schijndel, "Application of system identification methods to implement Comsol models into external simulation environments," 2009.

[7] F. Tjärnström, Variance Expressions and Model Reduction in System Identification. Department of Electrical Engineering, Linköpings universitet, 2002.

[8] W. Lai, M. Rahmat and N. A. Wahab, "MODELING AND CONTROLLER DESIGN OF PNEUMATIC ACTUATOR SYSTEM WITH CONTROL VALVE," in international journal on smart sensing and intelligent system, vol. 5, no. 3, pp. 624-644, 2012.

[9] P. Czop and G. Wszołek, "Advanced model structures applied to system identification of a servo-hydraulic test rig," Issues, vol. 1, pp. 2, 2010.

[10] L. Ljung, "MATLAB System Identification Toolbox User's Guide. The Mathworks," Inc., Sherborn, Massachusetts, 1986.

[11] B. Ninness and A. Wills, "An identification toolbox for profiling novel techniques," in 14th IFAC Symposium on System Identification, 2006, pp. 301-307.

[12] A. Wills, A. Mills and B. Ninness, "A matlab software environment for system identification," in Proceedings of the 15th IFAC Symposium on System Identification, St. Malo, France, 2009.

[13] Lathauwer L. Database for the identification of systems: FECG data EAST/SISTA KU leuven, belgium. 2008-02-22).[2013-02-8].Available: http://www.esat.kuleuven.ac.be/sista/daisy

Designing, Fabrication and Controlling Of Multipurpose3-DOF Robotic Arm

Hafiz Muhammad Nabeel, Anum Azher, Syed M Usman Ali and Abdul Wahab Mughal
Department of Electronic Engineering NED University of Engineering & Technology Karachi, Pakistan
Email: hafiz_nabeel91@hotmail.com, anum_azher@hotmail.com, ushah68@neduet.edu.pk, abdulwahabmughal@hotmail.com

Abstract. In the present work, we have successfully designed and developed a 3-DOF articulated Robotic Arm capable of performing typical industrial tasks such as painting or spraying, assembling and handling automobiles parts and etc., in resemblance to a human arm. The mechanical assembly is designed on SOLIDWORKS and aluminum grade 6061-T6 is used for its fabrication in order to reduce the structure weight. We have applied inverse kinematics to determine the joint angles, equations are fed into an efficient microcontroller ATMEGA16 which performs all the calculations to determine the joint angles on the basis of given coordinates to actuate the joints through motorized control. Good accuracy was obtained with quadrature optical encoders installed in each joint to achieve the desired position and a LabVIEW based GUI is designed to provide human machine interface.

Index Terms— Robotic Arm, Inverse kinematics, Microcontroller ATMEGA16 and Human machine interface (HMI).

1. INTRODUCTION

Nowadays industries are growing rapidly and they require solutions through which they can increase their productivity with precision and lower labor cost. Robotics is playing an important role in this aspect; it provides efficient, faster, safer and cost effective means to fulfill the industrial needs according to industrial standards. Robotic Arms are widely used all over the world; they came into existence in 1945 when da Vinci designed its first 4-DOF Robotic Arm with analog board for controlling [1]. First industrial Robotic Arm was made by Unimate in 1961; it has successively advanced into the PUMA Arm. Robotic Arms are extensively used in manufacturing and automobiles industries for welding, drilling, cutting, pick & place, quality assurance, assembling, safety and security, surgery, painting etc. to achieve high performance and precision. Depending upon the applications, Robotic Arms can be use in different configurations which include articulated (having at least three rotary joint), SCARA (having two parallel rotary joints), Cartesian (having three prismatic joints coincident with the Cartesian coordinates), anthropomorphic robot (similar to the human hand having fingers and thumb) [2]. We have designed an articulated Robotic Arm because it closely resembles human arm. As the articulated configuration contains all revolute joints, its movement is similar to human arm and can reach all the points within its work envelope. Autonomous electro-mechanical systems are more beneficial for the industries as compared to manual systems. A light weight, autonomous, computer-controlled Robotic Arm was developed by Lygorouas [3], a different approach was adopted by C. Flanagan for controlling the mobile Robot [4], and PIC16F877 microcontroller based system was designed by Hisham to control a 6-DOF articulated Robotic Arm [5]. Computer vision [6] technique is used to make a system autonomous where high precision and accuracy is required. However, those systems which do not require high precision and accuracy can be alternatively implemented using ultrasonic range finders [7].

In this work, we have designed and fabricated multipurpose Robotic Arm using ultrasonic range finders for object detection and obstacle avoidance instead of computer vision to provide a low cost industrial Robotic Arm. We have provided the flexibility to our designed Robotic Arm for having rotation of base and arms at 180 degree and 90 degree respectively. Designing of structure is done on "SOLIDWORKS" and Aluminum is used to make a light-weight manipulator. Inverse kinematic [9, 10] equations are used to determine the joint angle on the basis of given coordinates to actuate the joints. Three DC motors with quadrature optical encoders [8] are installed in each joint to achieve the desired position.

A microcontroller ATMEGA16 [12] is used to performs all the calculations based on the codes written in C-language using compiler AVR STUDIO. Inverse kinematic equations are fed into the controller to actuate the DC motors in order to reach the Robotic Arm at the required coordinates/positions in the work envelope. The driving circuitry based on IRFZ44N MOSFETS operates on the PWM techniques to drive the motors. We have also incorporated a Human Machine Interface (HMI) using "LABVIEW" to provide user friendly environment. The designed Robotic Arm can be used in industries for multiple applications such as material handling, automobile painting and etc.

2. DESIGNING AND FABRICATION

2.1 Mechanical Design

The mechanical structure was designed in such a way that it will have to support all the algorithms that will be applied on it. This mechanical design was successfully carried out using SOLIDWORKS. The specifications are given in the following table.

Parts	Maximum Angle (Degrees)	Speed (rpm)	Weight (Kg)
Rotating Base	180	$12 \pm 10\,\%$	4.08
Primary Arm	90	$(19 \pm 3\%)\,/36$	1.2872
Secondary Arm	90	$15.5 \pm 10\%$	1.292

TABLE I. MECHANICAL STRUCTURE SPECIFICATIONS

The payload of Robotic Arm is 0.5 Kg, its reach is 762mm (2.5 feet) and weight is nearly10 Kg.

2.1.1 CAD Designing

2.1.1.1 Base of the Arm The base consists of two parts; fixed and rotary. The function of fixed base is to hold first motor which causes the the rotation of base. The rotating base contains mounting of second motor and worm gear which increases the torque of second motor and reduces the speed of primary arm.

Figure1: CAD model of (i) fixed base (ii rotating base)

2.1.1.2 Primary Arm The upper part of the arm has smaller radius as compared to that of lower part. It was done so that point of gravity comes closer to the base. Mounting of third motor which holds the secondary arm is shown in Fig 2, square box showing space for the mounting of third motor, the coupling between the second motor and primary arm is also shown.

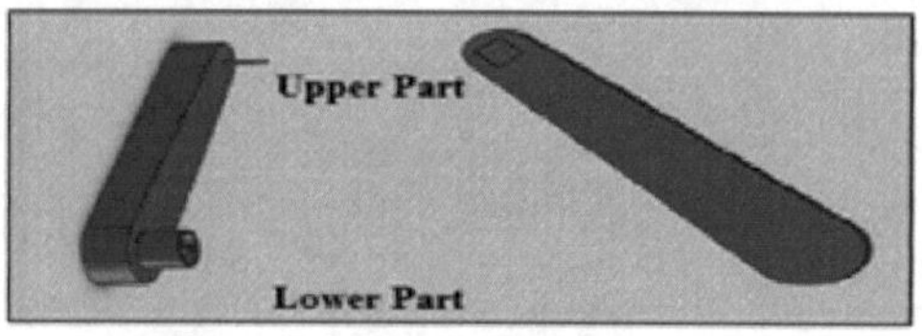

Figure 1: CAD model of Primary Arm

2.1.1.3 Secondary Arm The geometry of the secondary arm is similar to that of primary arm.

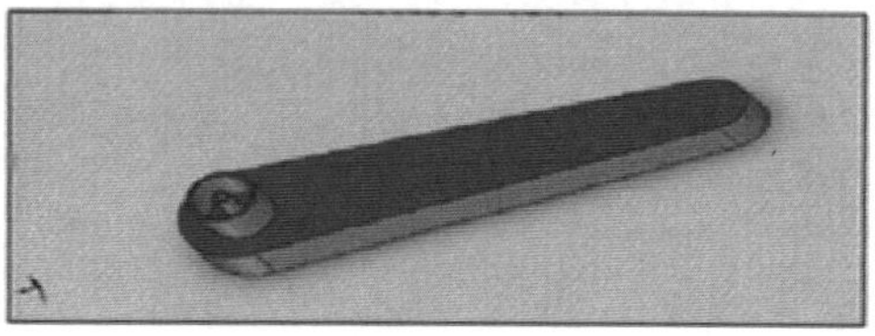

Figure 2: CAD model of Secondary Arm

2.1.2 End-effector A manual spray gun was automated by using a small DC motor. Since motor was not able to actuate the trigger of spray gun that's why a spring was used to aid the motor so that motor can easily actuate the trigger of spray gun.

2.1.3 Final Assembly

Figure 5: Final Assembly of the Robotic Arm

2.2 Motors selection

2.2.1 Selection of motors After studying about stepper, servo & DC motors, we have chosen DC motors with optical encoders as they are easy to use, easy to control and maintain as compare to others

2.2.2 Torque calculations for the motors selection Torque calculation has been done according to the geometry of the structure and SOLIDWORKS has been used to determine the required parameters for calculation which are not easier to find manually.

Total torque is given by,

$$\tau_m = \tau_g + I\alpha$$

Where, τ_m = Total torque, τ_g = Holding Torque

I= Moment of Inertia, α = Angular acceleration

$$\tau_{m1} = 2.96\ Nm\ ,\ \tau_{m2} = 13.245Nm\ ,\ \tau_{m3} = 3Nm$$

IOP Conf. Series: Materials Science and Engineering **51** (2013) 012023 doi:10.1088/1757-899X/51/1/012023

2.2.3 Specification of Selected Motors
Operating Voltage of motors= 12 volts
No load current =0.25A, 2.5A and 0.55A
Stall Torque= 30, 100, 30 Kg.cm
Since the minimum torque required for the second motor is 11.245Nm as per calculations but the purchased motor has the torque of 10 Nm therefore, a worm gear is used to achieve the desired torque.

2.3 Implementation of Forward and Inverse Kinematics

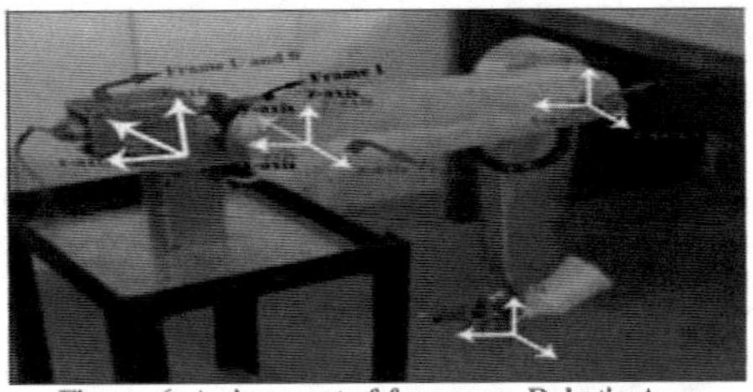
Figure 6: Assignment of frames on Robotic Arm

2.3.1 Transformation Matrix The general transformation matrix [9] is,

$$ {}_{i-1}^{\ i}T = \begin{bmatrix} c\theta_i & -s\theta_i & 0 & a_{(i-1)} \\ s\theta_i c\alpha_{(i-1)} & c\theta_i c\alpha_{(i-1)} & -s\alpha_{(i-1)} & -s\alpha_{(i-1)} d_i \\ s\theta_i s\alpha_{(i-1)} & c\theta_i s\alpha_{(i-1)} & c\alpha_{(i-1)} & c\alpha_{(i-1)} d_i \\ 0 & 0 & 0 & 1 \end{bmatrix} $$

Where, $s\theta_i = \sin(\theta_i)$, $c\theta_i = \cos(\theta_i)$

1. θ_i is the angle about z-axis
2. a is the length along x-axis
3. d is the length along z-axis
4. α is the angle about x-axis

Transformation matrix from U frame to 4 frame,

Total transformation from U frame to 4 can be obtained by multiplying all the matrices,
$uT^4 = uT^0 * {}_0T^1 * 1T^2 * 2T^3 * 3T^4$
$uT^4 =$

$$ \begin{bmatrix} c1c2c3 - c1s2s3 & -c1c2s3 - c1c3s2 & s1 & 1.7s1 - 3c1 - 16.922c1c2 + 25.5c1(c2s3 + c3s2) \\ c2c3s1 - s1s2s3 & -c2s1s3 - c3s1s2 & -c1 & -1.7c1 - 3s1 - 16.922c2s1 + 25.5s1(c2s3 + c3s2) \\ c2s3 + c3s2 & c2c3 - s2s3 & 0 & 1.3779 - 16.922s2 + 25.5(s2s3 - c2c3) \\ 0 & 0 & 0 & 1 \end{bmatrix} $$

2.3.2 Inverse Kinematics for base motor (θ_1): The inverse of u to 0 transformation matrix is taken for the calculation of θ_1,

$$ \boldsymbol{\theta_1 = \sin^{-1}(-1.7/\sqrt{Px^2 + Py^2}) + \tan^{-1}(Py/Px)} $$

2.3.3 Inverse Kinematics for secondary arm (θ

$$ \boldsymbol{\theta_3 = \sin^{-1}((936.6 - Px^2 - Py^2 - (Pz - 1.3779)^2 - 6Pys_1 - 6Pxc_1 9)/863.02)} $$

2.3.4 Inverse Kinematics for primary arm (θ_2): For calculation of θ_2, inverse kinematics as well as geometry has been used.

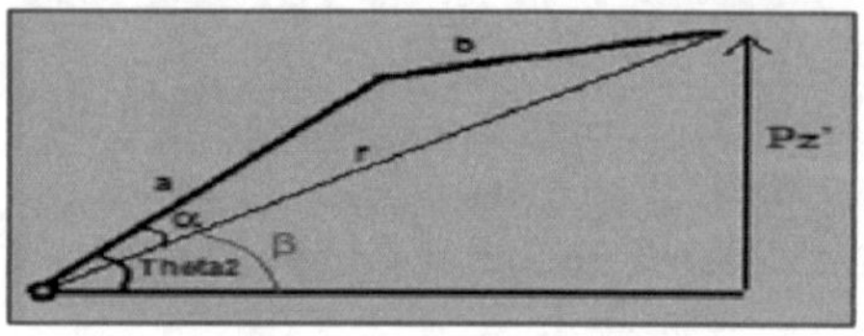

Figure 7: Diagram for calculating θ_2

$$\theta_2 = \sin^{-1}(Pz'/r) + \cos^{-1}((16.922^2 + r^2 - 25.5^2)/(2 * 16.922 * r))$$

2.4 Electronic Controller Board

2.4.1 Motor Driver Circuit An H-bridge circuit for controlling DC motor is designed. The circuit consists of four IRFZ44N, two IR2110 to drive the MOSFETs, Schmitt trigger circuitry for generating delayed and inverted PWM signal, diodes 1N5817 are used to provide the alternate path in case of back EMF and bootstrap capacitors are used to maintain V_{GS} of 12volt for each MOSFET.

Schmitt trigger 74hc14/7414
A single RC integrating circuit between the output and the input of an inverting Schmitt trigger produces a continuous square wave whose frequency depends on the values of R and C, and the threshold points of the Schmitt trigger.

Circuit has frequency range of 1KHZ to 50KHZ. RCD values are perfectly all right for this frequency range. The mentioned diode direction is used to produce the positive dead time and both channels need dead time not just one. If the diode direction is reversed then you will have a negative dead time which is not required.
Layout for motor driving circuit is designed by using software "Circuit Wizard" and FR4 material is used for fabrication of PCB.

2.4.2 ATMEL AVR Microcontroller Microprocessor and microcontroller act as the brain of the robot which tells it to perform tasks based on the inputs taken either from HMI or from the sensors. The overall operation of the robotic arm is controlled by programming the AVR microcontroller. Atmega16 [12] (8-bit microcontroller) is used because of its unique features including robustness, RISC architecture, high Endurance Non-volatile Memory segments, JTAG Standard Interface, Byte-oriented Two-wire Serial Interface, wide frequency range (0- 16 MHz), low power consumption and exclusive peripheral features.
The main features of Atmega16 which are used for the programming of robotic arm are:

- External Interrupts
- Three PWM channels
- USART
- EEPROM Data Memory

An LCD interface is also provided so that encoder count, joint angles or any other message can be displayed for the user.

2.4.3 LabVIEW Based GUI LabVIEW oriented GUI is developed to make the system user friendly. User enters coordinates via GUI and three motors of robotic arm will actuate so that end-effector achieve the desired coordinates in space. GUI contains two buttons, one for sending x, y and z-axis values to microcontroller and other for terminating program, and three blocks for taking x, y and z-axis.

3. CONCLUSION

In this work, we have successfully designed and fabricated a multipurpose articulated Robotic Arm having 3-DOF capable to perform various industrial tasks. By utilizing the Inverse kinematic equations the designed Robotic Arm has performed the all designated tasks assigned in a selected work envelop. We have also provided a user friendly environment using LabVIEW to control the functions of this manipulator by an unskilled labor or a technician in an industry. In future, image processing can be used to provide vision, and PID controller can be applied to develop a fully automated Robotic Arm.

References

[1] Michael E. Moran, "Evolution of robotic arms", J Robotic Surg(2007) 1:103-111
[2] http://prime.jsc.nasa.gov/ROV/types.html (last accessed Jan, 2012)
[3] John N. Lygorouas, Basil G. Mertzios, Nicholas C. Voulgaris, "Design and construction of a microcomputer controlled light-weight robot arm", Robotics and Autonomous Systems, Volume 7, Issue 4, November 1991, Pages 269-283
[4] D. J. F. Toal, C. Flanagan, "Pull to position, a different approach to the control of robot arms for mobile robots", Journal of Materials Processing Technology, Volume 123, Issue 3, 10 May 2002, Pages 393-398
[5] Ahmad M. Hisham, "A Computer Controlled 6-DOF Articulated Robotic Arm", RAS Newsletter – University of Waterloo, Issue 7, January 2009
[6] Mana Saedan and Marcelo H. Ang Jr.,"3d Vision Based Control on An Industrial Robot", Iasted International Conference On Robotics And Applications, Nov 19-22, 2001, Florida, USA
[7] J.M.P&ez Oria, A.M.Groba GonzBlez', "Object Recognition Using Ultrasonic Sensors in Robotic Applications", Universidad de Cantabria, Avda. de 10s Castros, sin. 39005 Santander
[8] http://mechatronics.mech.northwestern.edu/design_ref/sensors/encoders.html (last accessed March, 2012)
[9] Saeed B. Niku, "Introduction To Robotics", Prentice Hall, 2001
[10] John J. Craig, "Introduction to Robotics: Mechanics and Control", 3/E, Pearson Education India, 01-Sep-2008
[11] http://www.atmel.com/ (last accessed Oct, 2012)

IOP Conf. Series: Materials Science and Engineering **51** (2013) 012024 doi:10.1088/1757-899X/51/1/012024

Semantic Mapping and Motion Planning with Turtlebot Roomba

Rizwan Aslam Butt *, Syed M Usman Ali*

*Electronic Engineering Department, NED University of Engineering. & Technology, Karachi, Pakistan

rizwan.aslam@neduet.edu.pk, uashah68@neduet.edu.pk

Abstract. In this paper, we have successfully demonstrated the semantic mapping and motion planning experiments on Turtlebot Robot using Microsoft Kinect in ROS environment. Moreover, we have also performed the comparative studies on various sampling based motion planning algorithms with Turtlebot in Open Motion Planning Library. Our comparative analysis revealed that Expansive Space Trees (EST) surmounted all other approaches with respect to memory occupation and processing time. We have also tried to summarize the related concepts of autonomous robotics which we hope would be helpful for beginners.

1. Introduction

Motion planning is the core of autonomous robotics. The idea is fascinating that a robot manipulator is asked to reach an assigned position (goal) from its current position (start) in the workspace with the help of a motion planning algorithm, while avoiding hitting the obstacles in the way. Robot operating system (ROS) [1] is a quite popular open source development environment for robotic application development. Drivers and supporting packages are available for many robots in ROS like PR2 [2], Turtlebot [3] etc. Due to integration of Point Cloud Library (PCL) [4] and OpenCV [5] in ROS it is easier to develop visualization and navigation applications for ROS supported Robots. In this paper we have described our motion planning experiments on Turtlebot in our University Lab. We used Microsoft Kinect for vision and particle filter based Simultaneous and Localization And Mapping (SLAM) technique for generating grid based environmental map. Later on, we utilized this generated map for solving motion planning queries in ROS environment which were successfully solved by our robot. During the investigations, it has been observed that the performance of Robot was not consistent in solving the motion planning queries and the reason for such behavior was expected due to heavy computational load for working with complete map of the environment. Thus, based on our observed results, we decided to test the performance of our robot with sampling based motion planning approaches using Open Motion Planning Library (OMPL) by Kavraki lab [6]. This library provides the provision for testing as well as visualizing a motion planning algorithm. The state space and the robot model can be defined in collada file format. A motion planning query can be assigned using Scripting, Coding or through Graphical User Interface. We developed our environmental and robot models in collada file formats using Google Sketchup.

2. Related work

Autonomous and Assistive robotics is currently an active area of research. Especially ROS has gained wide popularity as a common open source platform for the development of perception and navigation activities using ROS supported robots like PR2, TOM ROSSIE [7] and Turtlebot. For example, Intelligent autonomous systems Group (IAS) has demonstrated autonomous robotic activities in kitchen environment in [8] [9] using PR2 and TUM Rosie. Similarly the Kivraki labs group is dedicatedly working on motion planning algorithms. Since it is hard to directly implement a Motion planning algorithms so it is often suitable to first test the algorithm in a simulation environment like OpenRave [10] and OMPL. As pointed out by Mark Moll at el. in [11] OMPL is a recent addition to such simulators through which it is easier to simulate and demonstrate motion planning scenarios without the need of going to complex and low level programming. The OMPL has now gained quite wide popularity in analyzing complex motion planning tasks. For example Christos Fragkopoulos et al has presented an analysis of various sampling based motion planning algorithms on 7 Degrees of Freedom (DOF) robotic arm mounted on the rehabilitation system FRIEND in a SE(3) configuration space. One more advantage of OMPL is that it can also be integrated with ROS via OMPL ROS interface. Therefore, we have chosen OMPL for analyzing various sampling based motion planning algorithms on Turtlebot robot.

3. Motion Planning with Turtlebot

Motion planning is the task for a robot of finding a collision free path from a starting point to a goal point and then actually following that path. This first requires some sort of semantic map of the environment like Grid based maps or Roadmaps and then locating its current position within that map. Both these tasks are pre-requisite for each other so also termed as chicken and egg problem. The solution to this problem is termed as SLAM. There were different SLAM solutions available to resolve such issue like [12][13] DP-SLAM, FastSlam , GridSlam , GMapping etc [14] . Each of these approaches is based on some of probabilistic estimator like Kalman filter, extended Kalman filter and particle filter. We used Gmapping package available in ROS which is a particle filter based SLAM technique.

3.1. Simultaneous Localization and Mapping (SLAM)

It is a famous technique used by autonomous vehicles and robots to develop not only a map of an anonymous environment but also to be able to keep track of their current position. If only we rely on the wheel encoder information for measurement of the current robot position in the map then the measured distance and direction travelled might be slightly inaccurate and thus the mapping process will be erroneous. Thus, It is required to improvise the erroneous measurements for the correct mapping otherwise earlier accumulate erroneous measurements will distort the whole map. Which means our measurements of robot's relative poses will also be erroneous. The solution for such issues is to have some static landmarks in the environment to improve our position measurements. SLAM approach provides us exactly the same mechanism. In SLAM process, we extract landmarks from the environment and improve our robot's position estimates with the help of some estimator. The typical example of such landmarks is walls in a room. For discovering landmarks in the environment we need a vision system like Laser Scanner, Ultrasonic sensors or Microsoft Kinect etc.

3.2. Semantic Mapping

is defined as the task of generating models of a robot's environment from sensor's data in which we have the information about the free space and obstacles present in this environment. In the perspective of indoor systems, three map concepts prevail namely; topological, geometric, and grids. Topological representations try to represent environments with structures like graphs, where nodes correspond to "something distinct" and edges represent an adjacency relationship between nodes. In cases of geometric models various geometric primitives are used for representing the environment. From the observation data the mapping is then responsible for estimating the best fit parameters of the primitives. Different representations, up till now, have been used with great success. One most successful primitive has been line segments used by many researchers as in [15] and [16] to represent parts of the environment. In Grid based approaches, the environment is divided in fine resolution grids with each marked as free or occupied. Occupancy grids are one of such popular approach which uses grid structures. In this approach the grid cells may be considered as pixels and the value of each pixel corresponds to the likelihood that its corresponding portion of workspace or configuration space is occupied [17]. Occupancy maps have been built with different approaches like stereo vision [19], Laser Scanners, and ultrasonic range-finders [18, 19, and 20].

3.3. Motion Planning Algorithms

There are different kinds of motion planning algorithms. Bug1 and Bug2 algorithms [21] are the oldest and simplest. These approaches treat the robot as a point in a plane with a contact sensor (or a zero range sensor) for detecting the obstacles. These techniques do not build any map of the environment and rather move along the obstacles like a bug to reach their destination. Their improved form is Tangential Bug [22] which used range sensor and thus has a global knowledge of the obstacles around and thus considered more robust. Another real-time method of obstacle avoidance and path planning was presented by Ossauma Khatib from Stanford University in [23]. It directs a robot from a start point to the end point as if it was a charged particle moving in a vector field gradient. Positive charge is assigned to the obstacles forming a repulsive force which directs the robot away from obstacles. The combination of attractive and repulsive forces hopefully guides the robot from the start location to the goal location while at the same time avoiding collision with obstacles. Another class of motion planning algorithms is mapping based planning. These methods require map of the environment in the configuration space and then considering the free spaces in the map as the nodes employ the graph theory concept for finding the shortest path from the robot's current pose to the destination pose. These methods often have two phases. In first phase, also termed as Global Planning, a shortest path algorithm like Dijkstra or A* is used for finding collision free poses to the goal point from the start point. In second phase a Local Planner algorithm is used to compute the control signals (velocity, acceleration etc.) suitable to drive the robot on the obtained collision free poses. As these map based methods require a lot of memory and computations therefore, it is more convenient to only consider some samples of free space instead of considering the whole leading to the class of algorithms termed as Sampling based Motion Planning Algorithms. These approaches can be broadly classified in three types; PRM (Probabilistic Roadmaps), EST (Expansive Space Trees) and RRT (Rapidly Exploring Random Trees).

4. Experimental Setup

Our experimental setup consisted of a Turtlebot Roomba with an iRobot Roomba base. For vision we had Microsoft Kinect mounted on it and an Acer Aspire1 notebook, with 1.66 GHZ atom processor and 2GB RAM, connected to it and placed within its racks as shown in Fig. 1. For driving the robot we used available ROS driver for Turtlebot. For teleportation we used the ROS keyboard package and kinect package to remotely control and view our robot.

Fig. 1. Showing picture of our Turtlebot Robot

4.1. Occupancy Grid Map of Radio Engineering Lab at our University

We used the Laser based SLAM implementation in ROS named as Gmapping. It is a particle filter based SLAM implementation. We configured it with 300 particles. Since Laser Scanner is quite costly and not available at NED University, we used Microsoft Kinect instead and used the ROS package Pointcloud_To_Laserscan [24] to convert the depth information of kinect to equivalent laser scan data. The idea was to read point cloud data of kinect at a certain height (10cm in our case) and read it at a certain spacing/linear angle. The intensity of the read pixels gave us distance information like a Laser beam in case of a Laser Scanner. We passed this data to PointCloud_Throttle nodelet which republished it to match the Laser Scanner publish frequency. Using this simple technique we were able to use our previous ROS packages and reproduce the real time semantic mapping of an indoor room environment as shown in Fig.5. The developed map was being viewed with the help of RVIZ [25] at the ROS master node, another laptop in our case. We saved this map using ROS map_saver [26] package to hard disk and is shown in Fig. 2.

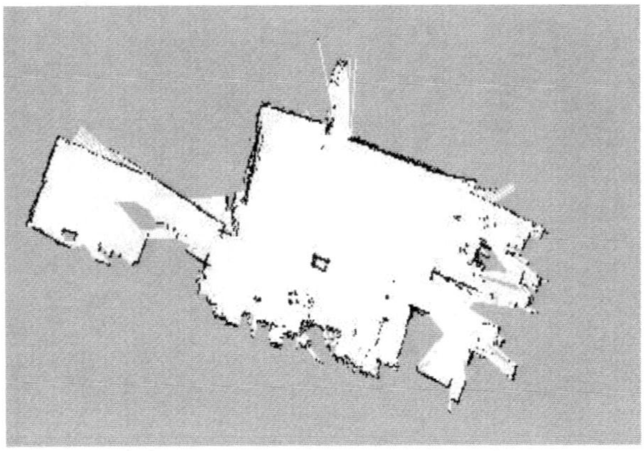

Fig. 2. Occupancy Grid Map of our University Lab

4.2. Motion Planning Experiments on Turtlebot

We used the obtained map to solve a motion query problem. We assigned a goal pose to our robot , used the ROS amcl [27] package to compute the current position in the map , navfn [27] package to obtain a global path in the map , which used Dijkstra's algorithm for finding shortest collision free path and used Trajectory rollout algorithm as a local planner available in nav_core [27] package. The performance of our robot was not consistent as expected and it failed sometimes. The reason for it was mainly heavy computational load on the notebook for considering the whole workspace.

4.3. Analysis with Sampling Based Approaches

As discussed that Sampling based motion planning approaches are faster and require lesser memory thus become more attractive for the researchers. To obtain best results, we have applied several approaches to our investigations in order to decide the best approach for our Turtlebot robot. We have performed the comparative study of EST , RRT , RRTConnect , LazyRRT , PRM , ERT , KPIECE , LBKPIECE , and SBL sampling based motion planning algorithms [30] using OMPL in an indoor closed environment .We developed required collada file format models of robot and environment in Google Sketchup. We selected a SE (3) environmental model from OMPL library resources and modified it. Fig.3 shows our experimental setup in the OMPL GUI. We used all the above algorithms one by one to solve this motion query with OMPL GUI by restricting bounds to allow the movement and rotation in x and y plane only , thus converting the problem to SE(2) instead of SE(3). The solved motion planning query with KPEICE showing the explored sample points is shown in Fig. 4.

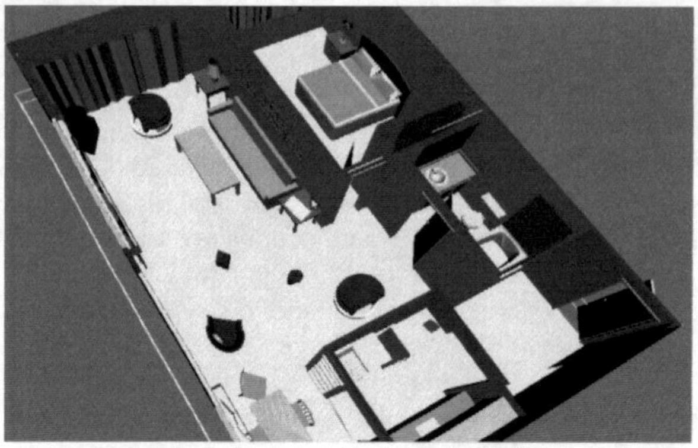

Fig. 3: Our Experimental setup in OMPL

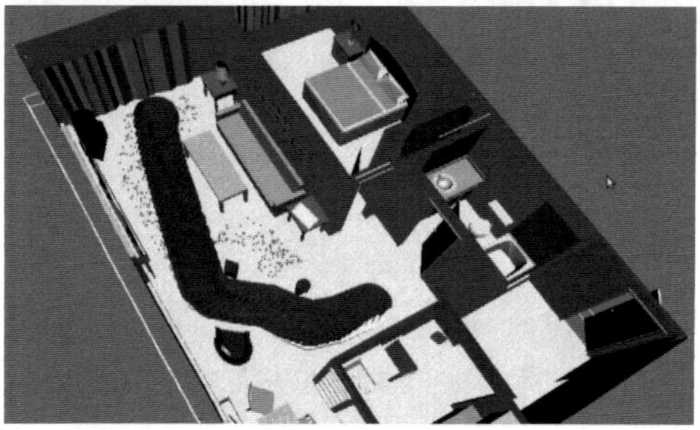

Fig. 4: Complete solved motion query result with Turtlebot using KPEICE

To elaborate the comparative study further we used the benchmarking utility of OMPL on the same experimental setup to compare all algorithm performance. We restricted the maximum memory usage for a planner to 2000MB. Our investigated results are shown in Fig. 5 (a) and (b). It can be seen clearly from these results that the PRM algorithm with uniform sampling approach consumed the minimal time to solve the motion planning query for our scenario but consumed a bit more memory but EST on the other hand had a slightly more simplification time but consumed the minimal memory. Therefore, EST proved to be the best motion planning algorithm for this scenario.

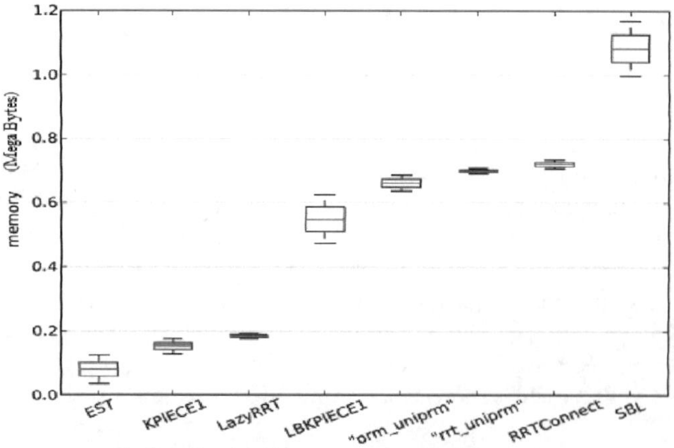

Fig. 5(a): Showing the memory consumption by each planner.

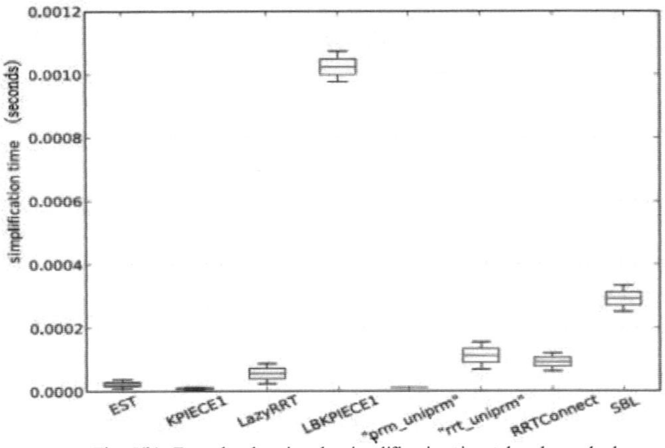

Fig. 5(b): Box plot showing the simplification time taken by each planner

4.4. Analysis with thin corridor

To analyze the complicated environment with our proposed system for solving motion planning queries, we introduced a thin corridor between the start and goal positions of our robot as shown in Fig. 6. The obtained results are shown in Fig. 7 (a) and (b) showing that the EST algorithm showed consistent and reproducible results with simple and complicated scenarios during several experiments.

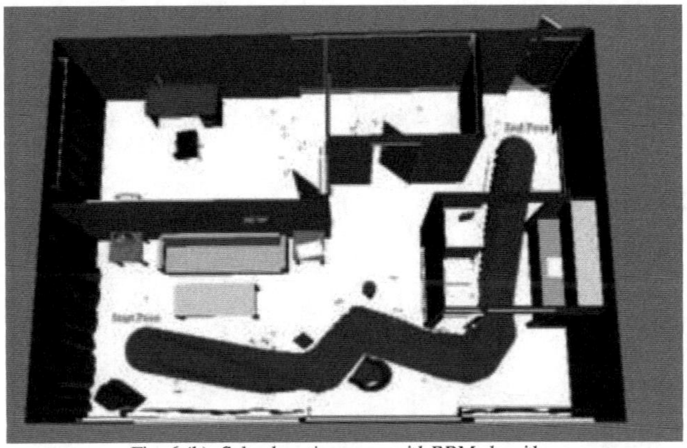

Fig. 6 (b): Solved motion query with PRM algorithm.

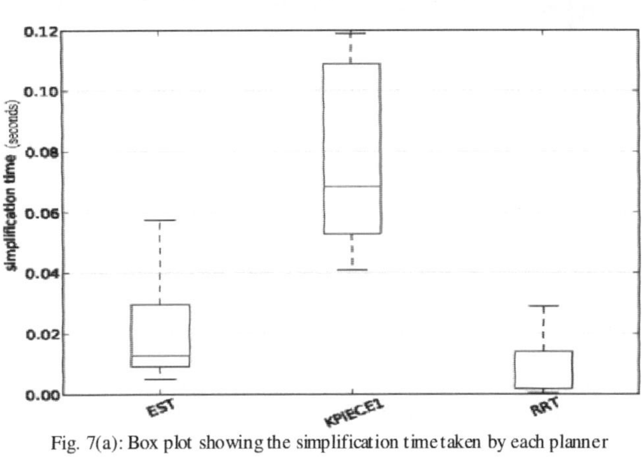

Fig. 7(a): Box plot showing the simplification time taken by each planner

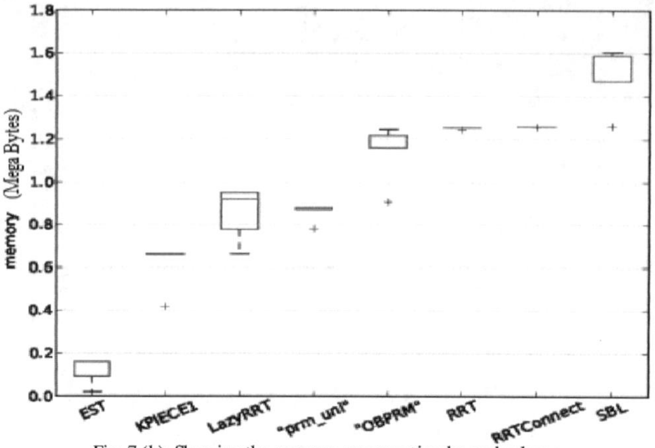

Fig. 7 (b): Showing the memory consumption by each planner.

5. Conclusion

The proposed developed system for implementing the EST based Motion Planning algorithm has successfully demonstrated the less memory consumption and processing time as compared to its other counterparts. The outcome from several experiments for simple to complicated scenarios during investigations produces the repeatable and reproducible results. Based on all these advantageous aspects of EST algorithm we conclude that it is most suitable sampling based motion planning algorithm for Turtlebot Roomba Robot that can also be suitable for any other planar robotic systems. Moreover, it is also possible to practically implement this algorithm on Turtlebot using OMPL_ROS interface.

References

[1] www.ros.org
[2] www.willowgarage.com/pages/pr2/overview
[3] http://turtlebot.com/
[4] http://pointclouds.org/
[5] http://opencv.org/
[6] http://ompl.kavrakilab.org/
[7] http://ias.cs.tum.edu/robots/tum-rosie
[8] Michael Beetz , IngoKresse, Ulrich Klank, Alexis Maldonado and others , "Robotic Roommates Making Pancakes ",2011, IEEE 11th International Conference on Humanoid Robots.
[9] Michael Beetz , Lucian , NicoBlodow, Cosmin Goron and others, "Autonomous Semantic Mapping for Robots Performing Everyday Manipulation Tasks in Kitchen Environments" , RSJ/IEEE International Conference on Intelligent Robots and Systems, 25-30 Sep 2011, San Francisco , USA.
[10] www.openrave.org
[11] Mark Moll, Janice Bordeaux , Ioan A. Sucan, Lydia E. Kavraki, "Teaching Motion Planning Concepts to Undergraduate Students ", IEEE Workshop on Advanced Robotics and Its Social Impacts (ARSO), OCT 2-4, 2011.
[12] Gamini Dissanayake , Shoudong Huang, Zhan Wang and Ravindra Ranasinghe , "A Review of Recent Developments in Simultaneous Localization and Mapping " , 6th International Conference on Industrial and Information Systems (ICIIS), Aug. 16-19, Sri Lanka.
[13] Durrant-whyte, Hugh, "Simultaneous localization and mapping: part I ", Robotics &

Automation Magazine, IEEE, Tim, Volume: 13, Issue: 2 pp: 99 - 110
[14] www.openslam.org
[15] Björn T. Jensen , Kai O. Arras, Nicola Tomatis and Roland Siegwart, "Multisensor on the fly localization: Precision and reliability for applications. Robotics and Autonomous Systems", vol 34, issue 2-3, page no 131–143, 2001.
[16] J. Crowley, "World modeling and position estimation for a mobile robot using ultrasound ranging " , IEEE International Conference on Robotics and Automation, 1989.
[17] A. Elfes , "Sonar-based real-world mapping and navigation. IEEE Journal of Robotics and Automation " ,vol 3, issue 3, page no. 249–265, June 1987.
[18] R.C. Gonzales (Eds.), Mongi A. Abidi, " Data Fusion in Robotics and Machine Intelligence ",Chapter 3, Academic Press, New York, 1992.
[19] D.R. Murray, J.J. Little, "Using real-time stereo vision for mobile robot navigation," Autonomous Robots, vol-8, issue 2, page no.161–17, 2000.
[20] G. Oriolo, M. Vendittelli, G. Ulivi, "Real-time map building and navigation for autonomous robots in unknown environments, " IEEE Transactions on Systems, Man, and Cybernetics, June 1998.
[21] V. Lumelsky and A. Stepanov , "Path planning strategies for point mobile automaton moving amidst unknown obstacles of arbitrary shape " , Algorithmica, vol-2, page no 403–430, 1987.
[22] Kamon, E. Rivlin and E. Rimon, ,"Tangent bug: A range-sensor based navigation algorithm" , International. Journal of Robotics Research, vol-17, issue 9, page no. 934–953, 1998.
[23] Oussama Khatib ,"Real Time Obstacle Avoidance for Manipulators and Mobile Robots" International Journal of Robotics Research,vol-5, 1986..
[24] http://www.ros.org/wiki/pointcloud_to_laserscan
[25] www.ros.org/wiki/rviz
[26] ros.org/wiki/map_server
[27] www.ros.org
[28] Ioan A. Sucan and Lydia E. Kavraki, "On the Implementation of Single Query Sampling Based Motion Planners ", IEEE International Conference on Robotics and Automation 3 to 7 May 2010

Wireless Control of Miniaturized Mobile Vehicle for Indoor Surveillance

Syed M Taha Saquib[1], Sarmad Hameed[1,3], Syed M Usman Ali[2], Raza Jafri[1], Imran Amin[3]
Department of Electronic Engineering, Usman Institute of Technology, Karachi-75300, Pakistan.
Department of Electronic Engineering, NED University of Engineering and Technology, Karachi-75270, Pakistan.
Department of Mechatronics, Shaheed Zulfikar Ali Bhutto Institute of Science and Technology,Karachi- 75600, Pakistan.
Email: taha.saquib@gmail.com, sarmad1987@gmail.com, usashah68@neduet.edu.pk , razajafri68@gmail.com, imran.amin@szabist.edu.pk

Abstract
This work is based upon electronic automation and Smart Control techniques, which constitute the basis of Control Area Network (CAN) and Personal Area Network (PAN). Bluetooth technology has been interfaced with a programmable controller to provide multi-dimensional vehicle control. A network is proposed which contains a remote, mobile host controller and an android operating system based mobile set (Client). The client communicates with a host controller through a Bluetooth device. The system incorporates duplex communication after successful confirmation between the host and the client; the android based mobile unit controls the vehicle through the Bluetooth module.

Index Terms — Miniaturized vehicle, wireless control, Bluetooth, programmable controller, duplex communication, android interface.

1. Introduction

The past decade can be termed as a revolutionary decade in the field of electronics; it provided new horizons for automation and control techniques. Various smart control techniques provided the designers to have more control over various applications. These techniques provided a great deal in control area network (CAN) and Personal Area network (PAN). A secure, accurate and effective remote surveillance monitoring is a crucial requirement these days.

Modern communication interfaces protocols include, WIFI, ZIGBEE, and Bluetooth etc. Each of the connection has their own unique specifications and applications. The comparisons of the protocols are listed in table 1.

	Zigbee	Wi-Fi	Bluetooth
Range	10-100 Meters	50-100 meters	10- 100 Meters
Networking Topology	Ad-hoc, peer to peer, Star, Mesh	Point to hub	Ad-Hoc, very small networks
Operating Frequency	868 MHz (Europe) 900-928 MHz (NA), 2.4 GHz (worldwide)	2.4 and 5 GHz	2.4 GHz
Power Consumption (Battery option and life)	Very low	High	Medium
Security	128 AES plus application layer security	Encryption and MAC address	64 and 128 bit encryption
Typical Applications	Industrial control and monitoring, sensor networks, building automation, home control and automation, toys, games.	Wireless LAN connectivity, broadband Internet access	Wireless connectivity between devices such as phones, PDA, laptops, headsets(smart Phones)
Android Connectivity	Not possible	Possible (not Possible)	Direct Connection is possible

Table1: Comparisons among various techniques.

Among the three popular wireless connections Bluetooth has been chosen with its suitable capability. Bluetooth a short-range communications technology that is simple, secure, with globally available frequencies of 2.4GHz is able to provide connectivity up to 100 meters at speed of up to 3Mbps depending on the Bluetooth device class [1]. In addition, a Bluetooth master device is able to connect up to 7 devices in a "Pico net" [2].

Bluetooth communication has been interfaced with android operating system which is a Linux-based operating system designed primarily for touch-screen mobile devices such as smart-phones and tablet computers. It is open source software, hence cost effective.

In this paper, the proposed navigation method of the mobile surveillance system is evaluated in small premises.

2. System Description

The Bluetooth based mobile surveillance vehicle (MSV) design architecture, proposes a Bluetooth based MSV that is controlled by an android mobile's GUI, but it does not provide remote function for a larger coverage. For the system, all the controls are performed only at the GUI on android operating system (OS) based mobile phone is designed with cellular phone remote control. The system configuration is shown in Fig # 1.

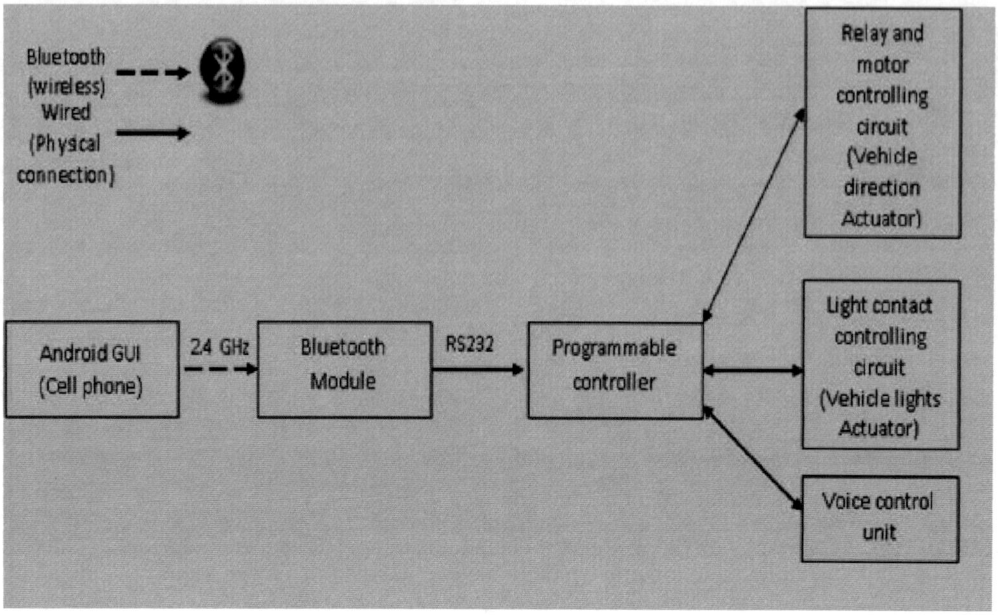

Figure-1: block diagram of the system.

A mobile phone with an android based operating system is proposed to control the mobile surveillance vehicle (MSV)'s devices and actuators wirelessly by means of Bluetooth which gives easy access for communication within an indoor proximity of 100 meters. The Bluetooth module is further connected with programmable controller via serial communication (RS232). The programmable controller performs the actuation on signal received from Bluetooth communication channel by making the vehicle perform its desired task. The controller is programmed to record the surrounding sounds. The controller is capable to establish duplex communication with the actuators thus getting feedback if any error occurs in the operation. The errors in the system are continuously monitored so that the efficiency of the system can be increased with minimum error/ loss rate.

3. Hard ware Design

The Android application on mobile phone sends the character to Bluetooth module which forwards the character to Ardunio microcontroller. The Arduino microcontroller receives that character as a numerical value and then matching the condition sends pulse on the desired port. The Pulse will continue to transmit until the button on android application is in contact.

Arduino microcontroller is being used in the project, which is an open-source single-board microcontroller, descendant of the open-source Wiring platform and designed to make the process of using electronics in multidisciplinary projects more accessible.

The hardware consists of a simple open hardware design for the Arduino board with an Atmel AVR processor and on-board input/output support. The software consists of a standard programming language compiler and the boot loader that runs on the board.

Bluetooth module is the most reasonable and easiest way to go wireless for this module, the 4 pins are +5V, GND, TXD, RXD. Supply voltage should be 3.3 - 6 V. Absolute maximum is 12V.
The connections for duplex communication channel of programmable controller are displayed in fig#2.

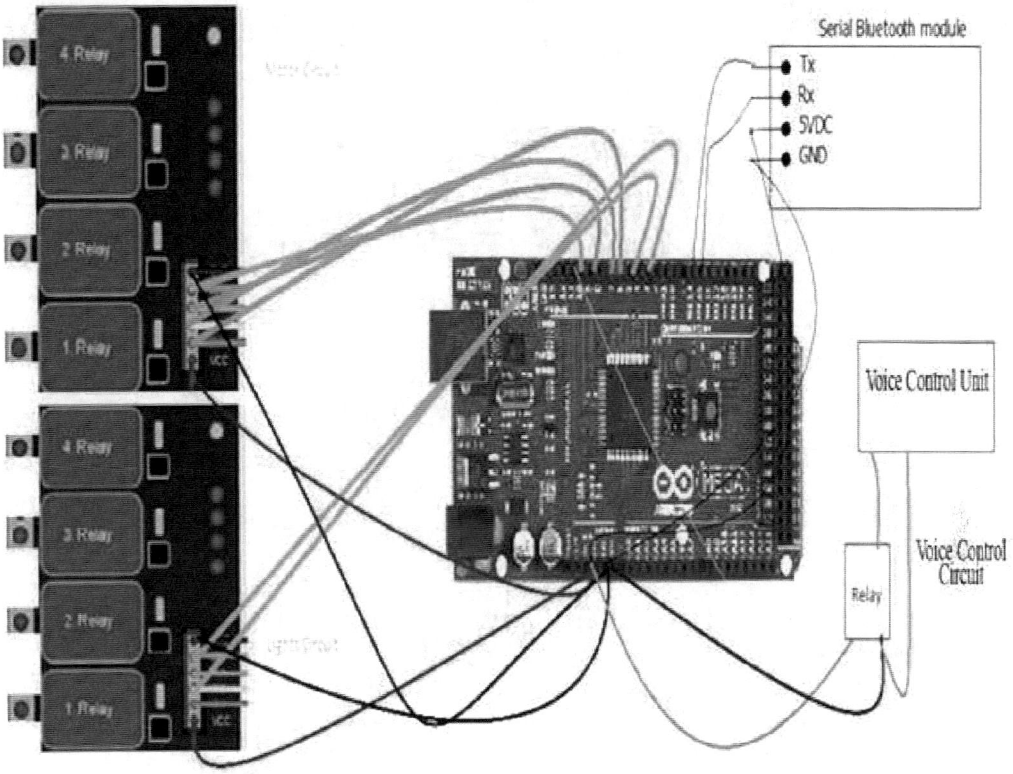

Figure # 2: pictorial for connection with Arduino microcontroller

The relay connection circuit is further listed in figure # 3. There are two modules of these Circuits, one for mobile surveillance vehicle (MSV)'s movement direction and another for front and backward motion control. The vehicle can perform basic movement control via forward, reverse, left and right turns with 360° rotation with two front and back lights.

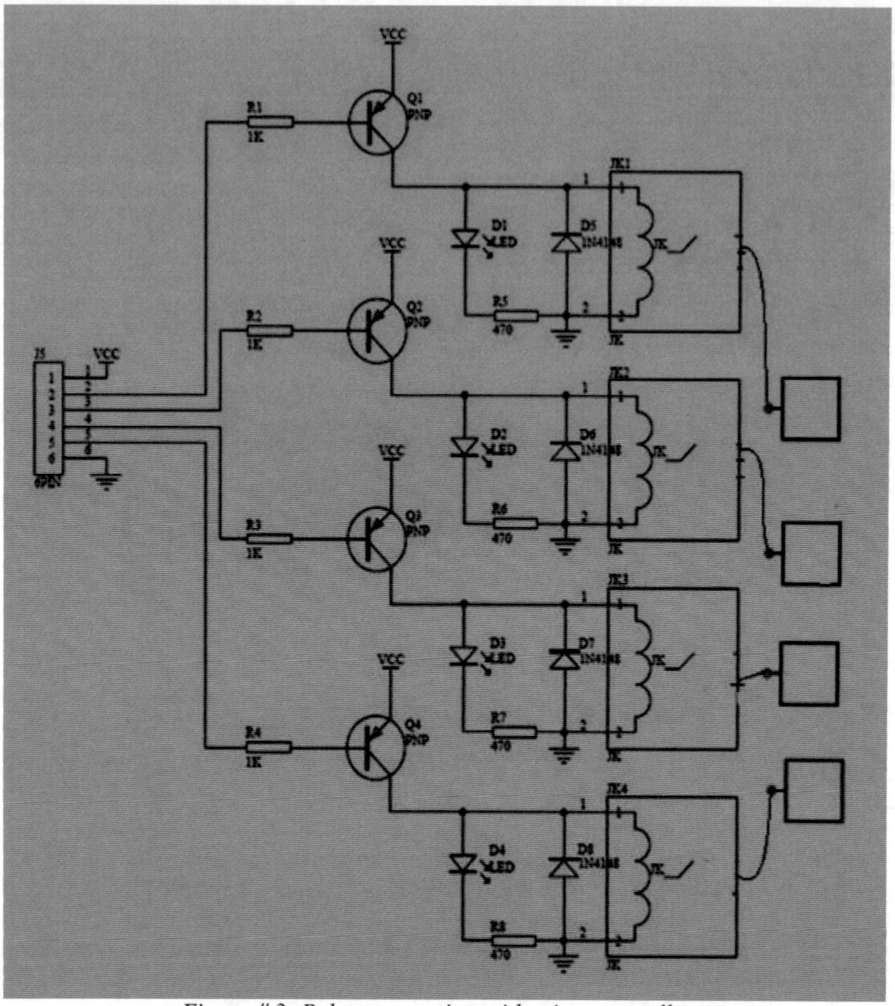

Figure # 3: Relay connection with microcontroller.

4. Soft ware Design

The software design mainly revolves around the programming of the microcontroller. The Arduino microcontroller is an open-source single-board microcontroller which is programmed in C or C++. Figure 4 illustrates the process of activating various controls in the system. The command detection function is performed by the microcontroller. The Android application on mobile phone sends the character to Bluetooth module which forwards the character to Ardunio microcontroller. The Arduino microcontroller receives that character as a numerical value and then matching the condition send pulse on the desired port.

The Pulse will continuously be transmitted as long as the button on the android application remains depressed. The activating actuators are designed to react to this pulse signal. If any of the input switches is released; it will interrupt the main function loop of the microcontroller. Then, the microcontroller will deactivate the relay and stop the ongoing task.

The proposed system decision command flow is further listed in figure#4.

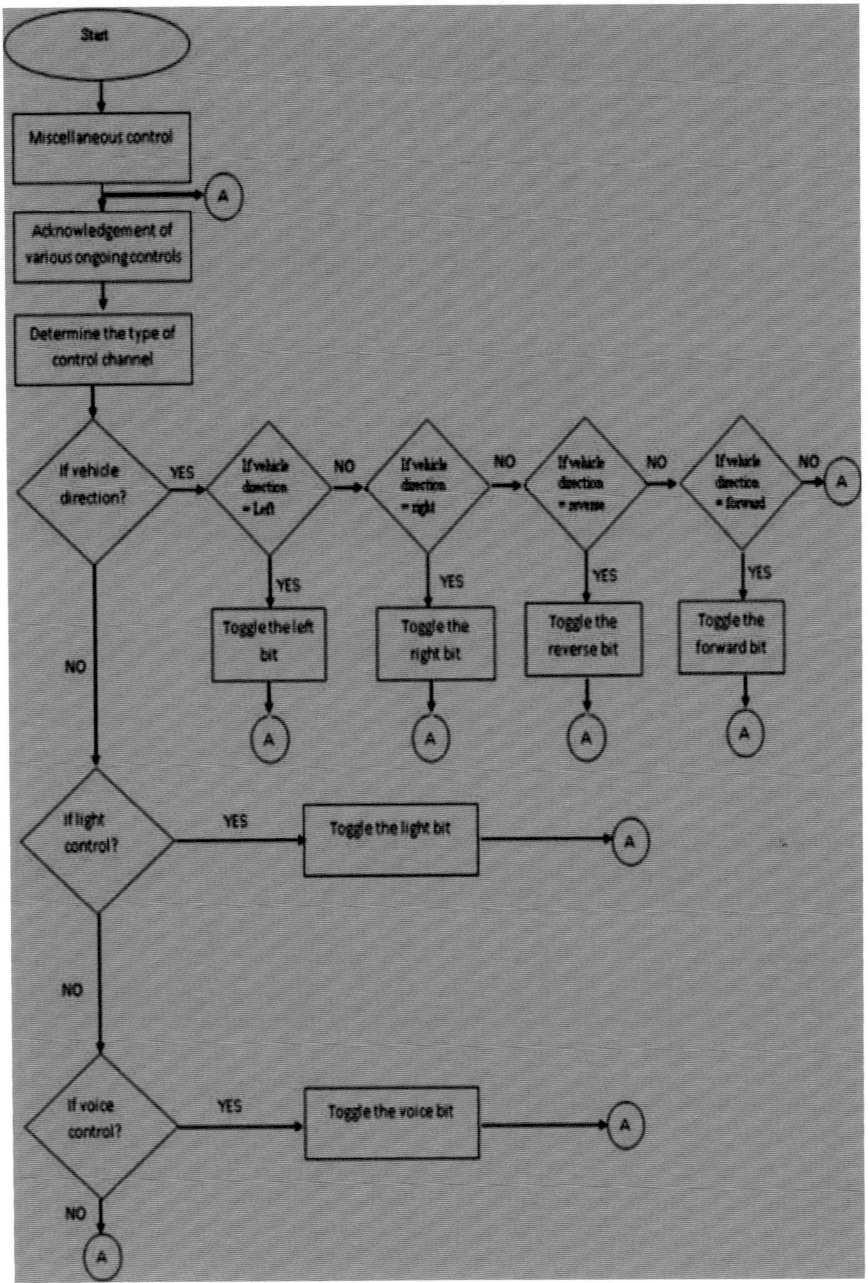

Figure #4: System decision command flow

Android GUI is designed by a user-friendly interface. The four directions indicate the desired control for the MSV that is connected to the system. User can simply click on the direction button to toggle the various controls. Control board's "Connect to BT"" button is provided to establish connection to main control board by Bluetooth. Phone's "Connect to BT"" button establishes connection between Android GUI and Bluetooth module. When the connection is established, Android GUI acts as the control panel between main board and the phone. All the data received from main board is forwarded to the phone. The data sent from the phone is also directed to the main board.

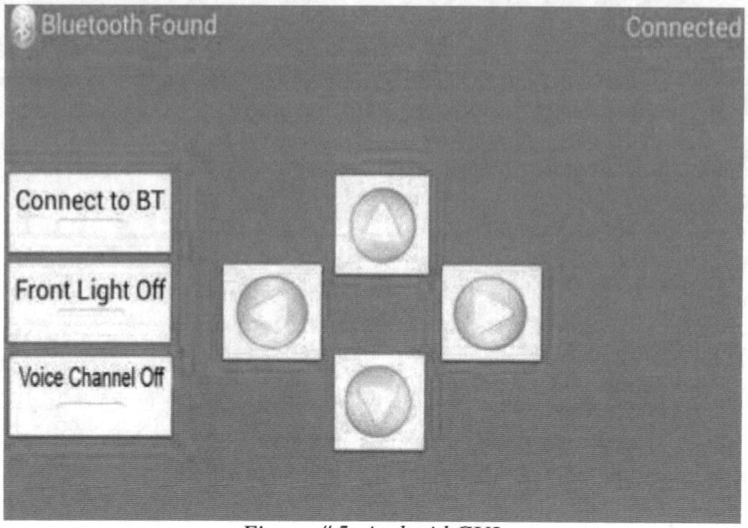

Figure # 5: Android GUI

The application is designed in low API level so that the devices with higher version are compatible with it. Fig 4 illustrates the Android GUI tested on smart phone with android Version 4.0.4 (Ice Cream Sandwich). The interface is simple to use, user can simply touch the icons on the interface to perform MSV's control tasks after been connected to the Android GUI.

5. Conclusion

A simple, low cost, user friendly and effective control system for MSV has been implemented using the android based SMART phone and the Bluetooth technology that is ideal for tactical remote surveillance and monitoring.
The connected GUI is synchronized with the control board to indicate the real-time control action implementation feedback.

For future work, the android GUI could be implemented with voice control. The android based GUI could also be replaced with windows based GUI for making the system compatible with non-android based applications. The system can also be equipped with a wireless camera that transmits the real time images. The vehicle can also be replaced by an intelligent remote sensing robot with decision making capabilities.

Acknowledgment
The authors extend their gratitude to their parent institutions that created a research based environment for continuing with projects like this and providing the necessary motivation for such activities.

REFERENCES

[1] The official Bluetooth website from Bluetooth SIG: *http://www.bluetooth.com*
[2] J. *Mander*and D. Picopoulos, "Bluetooth Piconet Applications," pp. 1-25.
[3] N. Sriskanthan and Tan Karande, "Bluetooth Based Home Automation Systems," Journal of Microprocessors and Microsystems, 2002, Vol. 26, pp. 281-289
[4] Wijetunge S.P., Wijetunge U.S., Peiris G.R.V, Aluthgedara C.S. &Samarasinghe A.T.L.K., "Design and Implementation of a Bluetooth based General Purpose Controlling Module", in IEEE, 2008, pp. 206-211

IOP Conf. Series: Materials Science and Engineering **51** (2013) 012026 doi:10.1088/1757-899X/51/1/012026

Energy Efficient and QoS sensitive Routing Protocol for Ad Hoc Networks

Tariq Saeed Tanoli[1], **Muhammad Khalid Khan**[2]
College of Computing & Information Sciences
Karachi Institute of Economics and Technology
Karachi, Pakistan
tariq@pafkiet.edu.pk[1], Khalid.khan@pafkiet.edu.pk[2]

Abstract. Efficient routing is an important part of wireless ad hoc networks. Since in ad hoc networks we have limited resources, there are many limitations like bandwidth, battery consumption, and processing cycle etc. Reliability is also necessary since there is no allowance for invalid or incomplete information (and expired data is useless). There are various protocols that perform routing by considering one parameter but ignoring other parameters. In this paper we present a protocol that finds route on the basis of bandwidth, energy and mobility of the nodes participating in the communication.

Index Terms: Adhoc Networks, Reactive Routing, QoS Routing, Bandwidth Reservation, FPGA

1. Introduction

Although MANET provides information through unguided media, there is no mandatory organized communication needed. They can be deployed anywhere at any time. The mobile nodes in MANET are self cooperating and they act as a router. Since the topology in MANET changes frequently due to the mobile nature of the node, the traditional protocols of routing are not efficient. Routing Protocols can be classified into two main types Proactive and Reactive.

Proactive protocols share network topology information and monitor the topology changes. Reactive protocols are on-demand routing protocols that find a route when it is necessary and save the route information for future use.

As WAN requires efficient battery power utilization, our designed protocol can be very useful in these scenarios, due to the overhead produced by the control packets. If it is required that there is no motion among devices, spontaneous protocols can be exercised. Using Hybrid protocol nodes can change their behaviour from reactive to proactive and the other way round, to save their battery power and to tackle mobility.

One feature of the practical routing protocol is that they transmit the information uniformly in order to check updates in the topology if any occurs and that information can be used for successful transmission with low delay. This type of routing protocol can have two types. The position of nodes plays very vital role to maintain topology, as nodes are mobile therefore there is strong need to send information to every mobile node including position of the other nodes and for that principle a lot of battery will be required. This high requirement of the battery is the one of the drawback of the pro-active protocols.

Therefore, to resolve the above issue, we need other routing protocols which can use the battery more effectively. Spontaneous protocols (Reactive) is one approach to solve the above mentioned problem, it connects the nodes only at the time of the requirement (or when we say there is data to send between the nodes by using any procedure contact between nodes). On one hand, the issue of battery life is solved, because now no connection is made when nodes have no information to send. But, on the other hand this practice has one drawback in cases where the network contains more than one hop. When there are two or more devices entertained in any network using spontaneous protocol, they should be able to create link between several nodes at the time of demand else searching for link could be the reason to effect the overall bandwidth as well as battery power.

2. RELATED WORK

[1]Cross layer optimization between MAC and network layer for routing. For a given path between a source and a destination, each intermediate node computes a new maximum number of transmissions based on a specific algorithm [1]. This parameter can be adjusted easily by each node. Using this new routing, we achieve a better average delay (resp. throughput) for each connection without changing the average throughput (resp. delay). In extreme cases, a reset technique is introduced to optimize performances [1].

The two layers are clearly separated. Attempting the channel begins by choosing the queue from which a packet must be selected. And then, this packet is moved from the corresponding queue from the network layer to the MAC layer where it will be transmitted and retransmitted, if needed, until its success or drop. In this manner, when a packet is in the MAC layer, it is itself attempted successively until it is removed from the node [1].

[2] In this paper writers have designed a protocol to increase the power efficiency. That protocol is the combined approach of both spontaneous (Re-active) protocol and practical (Pro-active) protocol. Based on the destination point, portable nature of nodes and remaining power, issue of battery life is solved. The author has focused on one protocol that is the ODMRP (ON-Demand Multicast Routing Protocol), which helped to know the spontaneous activities among the nodes that are necessary to create the links. The above mentioned protocol is the onward approach of the OLSR, in which every node has the authority to calculate its remaining power, remaining energy and mobile behavior. In the ODMRP, it is required to report to the beside node about any changing in the topology for that reason broadcasting is used.

[3] ORRP finds the shortest path without broadcasting the control packets by sending initiated protocol it assumes that at any given instance, any node in the network maintains a list of its neighbors and also stores the cost vectors to reach the neighboring nodes from the node [3]. Any change in the topology including deletion of a host or a link must be communicated to the neighboring nodes by an algorithm [3].The procedure update assumes that each node maintains a list of neighbors. A cost vector stores the costs of reaching the members of [3] it stores the cost of reaching node from N.

[4] To reduce the effect of interference, the low connection obstruction between transmitter and receiver can be used. In this paper the term average connection obstruction is utilized, instead of the low or high connection obstruction, because these both terms can change the actual measurement.
The protocol designed in this paper is the ALIR, it calculates the obstruction between the nodes and attempt to sort out the reason behind it. After the specific mechanism applied finally it is possible to find the average obstruction between transmitter and receiver. The RRP namely route reply packet is send from source to destination that is helpful to find the actual information about the interference in the connection. When the destination receives this RRP, based upon this packet it is easy to find the average connection obstruction. This information can be used to update the routing table in which the path between sources to destinations is included. The nodes in between transmitter and receiver will also revise their required tables in order to forward the data to the proper destination.

[5] This paper is based on the departed node issue or in another wording when destination is not available, to solve this issue without wasting efficiency of the routing protocols. The several topologies and problems geographic location service and greedy base algorithm solve the problem of dead end problem for wireless routing topologies by following a mechanism and create optimal connection between devices that is selected randomly using selection algorithm. For that purpose there is requirement of the to share the node ids and broadcast the initial packet but by following above mechanism we would have greedy routing algorithm limitation dead end problem periodic adjustments of hop-id causes the duplications of hop-id.

[6] In this paper IQRP (Interference aware QoS routing protocol) protocol is introduced. It optimizes the network resource utilization. The term 'LAB'' is discussed in this paper, if the value of this term is higher on any node it means that node has the low interference and is known as the practicable for the communication in any connection between the source and destination. It has got application in the both one-hop and two-hop neighbor conditions.

3. PROTOCOL DESCRIPTION

In our protocols some nodes are in proactive state and rests are in reactive state. When a node change its state it broadcast **STATE PACKET** on the network nodes having the proactive state will listen these packets and reactive will discard these packets. A node on leaving a group will be broadcast **QUIT PACKET** on the network these packets will again discard by reactive nodes of the network .To join a network applicant node broadcast a **JOIN PACKET** only proactive node will reply to those packet by sending a **ACK PACKET** to the applicant node then applicant node broadcast its **STATE PACKET** on the network as describe above on the description .If a node changes its state more than 3 times in X seconds so the proactive nodes mark that node as bad node and stop listening the broadcast from the that particular state and start **HOLD-DOWN TIMER** till Y seconds after sending the timer proactive nodes start listening the packet from bad node by resetting the **STATE TIMER**.

- X= Z / 3 (X is STATE TIMER)
- Y= Z / 2 (Y is HOLD DOWN TIMER)

Where Z is the Convergence time of the network.

Mobility Consideration

In our propose mechanism if a node change its location so it sends the **QUIT PACKET** on the network and all the proactive node discard the entry of this node when the node covers its distance and wants to join the network so it sends the **JOIN PACKET** on the network and proactive node

IOP Conf. Series: Materials Science and Engineering **51** (2013) 012026 doi:10.1088/1757-899X/51/1/012026
reply with **ACK PACKET** and then the requested node broadcast it **STATE PACKET**. If a node changes its state more than 3 times in X seconds so the proactive nodes mark that node as bad node and stop listening the broadcast from the that particular state and start **HOLD-DOWN TIMER** till Y seconds after sending the timer proactive nodes start listening the packet from bad node by resetting the **STATE TIMER**.

4. BANDWIDTH CONSIDERATION

The widest shortest path provided by Algorithm to provide the requested Bandwidth, all nodes whose local available bandwidth is smaller than the requested bandwidth and all nodes in the interference area of such transmitter nodes are temporarily removed from the one and two hops neighborhood and the topology table. [7]

Proof: We prove this property by induction on the path length. By construction, this property is true for a path length = 1 and also for 2. We assume that this property is true until rank n − 1 and we prove it at rank n. Let us consider the shortest path A0, A1, ...,An between A0 and An with n ≥ 2 meeting the bandwidth request Bf . It follows that the path A0, A1, ...,An−1 of length n − 1 is a shortest path between A0 and An−1 meeting the bandwidth request Bf . As the property is true at rank n − 1, Algorithm 1 provides the widest shortest path between A0 and An−1 meeting the bandwidth request Bf, let A0,A_1, ...A_n−2,An−1 be this path. As A_n−2 is one-hop from An−1 and An−1 is one-hop from An, A_n−2 is two-hop from An. Let A_n−1 be the MPR of An allowing to reach A_n−2 with the highest bandwidth. As An−1 allows to reach A_n−2 by a path meeting Bf , the MPR selected by An meets Bf . Algorithm will find the path A0,A_1, ...A_n−2,A_n−1,An.[7]This path is a widest shortest path between A0 and An, meeting Bf. After calculating the best path the senders mark the 1 to the route and then look for the required Energy.

5. ENERGY CONSIDERATION

Before presenting the protocol specification for Energy, we describe the Pricing scheme since the choice of the pricing scheme determines the minimum amount of Information that must be communicated to the destination node (which is in charge of computing the payments). Any pricing scheme that achieves individual rationality, truthfulness, and energy efficiency and pays only the nodes in the winning path must be based on the VCG mechanism. When adapted to our setting, the VCG mechanism, which optimizes the socially desirable goal of energy competence, it contains the certain laws through which optimal path and other parameter are selected. Whereas c(p) represents the cost and S & D are just source and destination of the connection.

$$c(P) = \sum \quad l(v)$$

The terms MP and v represent the minimum path cost and successful connection respectively. And c(p^-v) is the cost between the source and destination. If there is no any successful connection it means then path and winning path will be low. The final payment for any successful connection can be found by using below equation

$$pay(v) = c(P^{-v}) - c(MP) + l(v)$$

6. APLICATION AND MOTIVATION

Assume a Network of multi mobile nodes some of them are in proactive state and rest of them are in reactive state. Proactive nodes maintain their routing table and reactive maintains on demand temporary routing table.

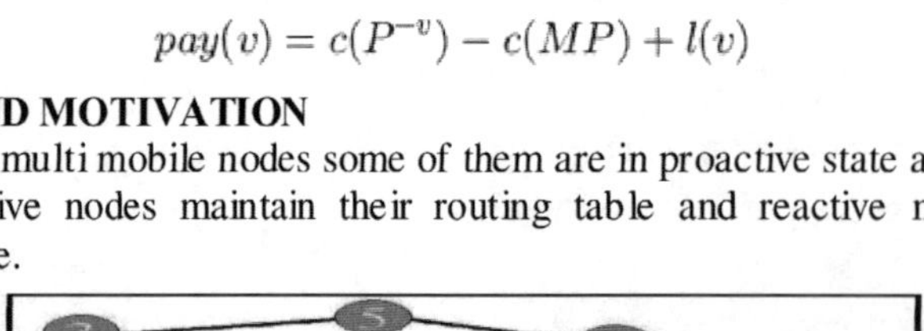

Figure 1

In Figure 1 Node 4, 7, 2,11,8 are in reactive state and rests of them are in proactive state and maintain their routing table. Now we suppose the node 3 want to send the data to node 9 hence it's a proactive node so its already maintain its routing table through above algorithm of Energy and Bandwidth and the table is maintain in (Figure 2).Node 3 has three different paths to sends data to node 9 .Node 3 compute the costs of all three path by energy and Bandwidth mechanism and find out the route best between source to destination is 2, 1, 6.

If the selected path energy and Bandwidth is enough to send data over the network so the sender marks it 1 and sender forward the packet on the route which has the maximum cost. The cost of route will be calculated on the basis of 0 and 1 the best path will always be 2 by adding the bandwidth and Energy cost and the next best path will be 1 to transfer a packet on the network. Figure 2 is the routing table of a Node 3 which has the cost of 2 and the rest of path will be the backup route which will be use in case of failure of any node or mobility nature of any intermediate node. Reactive nodes also maintain the same type of temporary table but they will maintain this when the require (On demand Routing).

Source	Destination	Intermediate	Bandwidth	Energy	Cost
3	5	4	1	1	2
3	9	5,4,7,6,10	1	0	1
3	9	5,8,11,10	0	1	1
3	9	2,1,6	1	1	2
3	4	2	1	1	2
3	11	2,4,7	1	1	2
3	7	2,4	1	0	1

Figure 2

If any of intermediate, source or destination node change its state the transfer would not effect and if the node change its position so it will broadcast the quite message on the network and when it needs to join the network it sends the join request to the network and proactive nodes will reply to the requested nodes. suppose node 8 is fake node and change its state continuously and sends state packet on the network so the proactive node waits till state timer X then declare this node as fake node and stop listening its broadcast will hold down timer Z. the computation of these timer has already been discussed in the paper.

7. ASSUMPTIONS

For implementing the protocols we have five nodes three of them are in reactive state and other two are in proactive state. Each link has the 1 MB of data transfer rate and each link utilize its whole bandwidth to transfer the data on the network with 10 microseconds of delay. Initially node 0 ,1 ,4 are communicating with node 2 and node 2 is communicating with node 3 and acting as a bridge for node 3 to communicate on the network though packets may be dropped with certain probability based on the configuration. This simple model enables us to abstract the impact of message loss on routing performance and allows us focus on how well the proposed routing mechanism performs. .following is simulation results of our proposed protocol are shown below (NAM View).

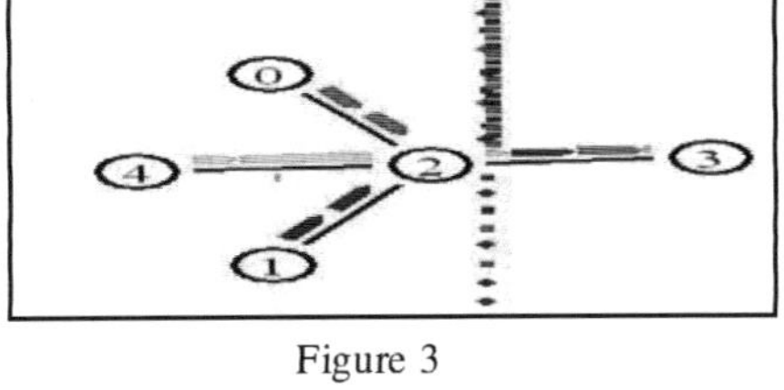

Figure 3

A. Simulation Results

Figure3 simulation is the propose mechanism of our protocol, the mobile nodes are communicating with each other, node 0, node 1 and node 4 are transmitting their data to node 2 and because of convergence at node 2 all the data are in queue and after filling the queue the remaining packets would be dropped at node 2 and in the mean time node 2 is also communicating with node 3.

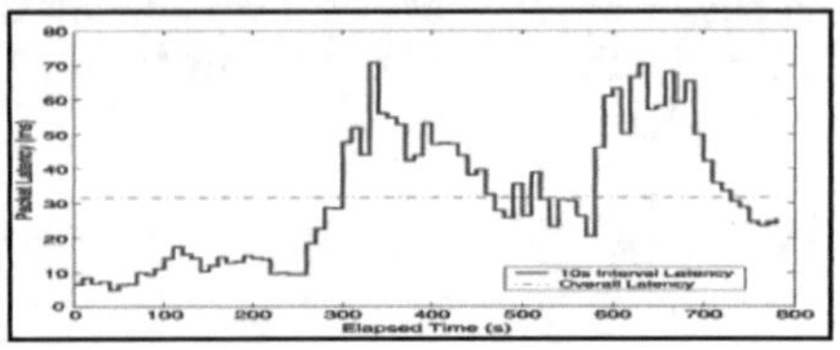

Figure 4

The figure4 shows the protocols performance as a function of the mobility speed. Figure4 shows the packet latency along with elapsed time.

B. Benefits of Protocol

Our proposed protocol is capable of finding a path on the basis of energy, bandwidth and mobility nature of nodes along with QoS by maintaining the alternative path of the destination

8. CONCLUSION AND FUTURE WORK

In this paper we have shown how to provide QoS routing on the basis of Energy, Bandwidth and mobility nature of nodes and restrict the fake node on the network to avoid flooding.

High computation of Energy and Bandwidth would drain the battery and increase delay produced by the network convergence. In large network where we have large number of nodes on the network so each proactive node has to maintain that much large routing table and it would be possible after some time because of energy the proactive node change it state to reactive.

9. References

[1] Dynamics Retransmissions Limit Scheme for Routing in ulti-hop Ad hoc Networks alph El Khoury and Rachid El-Azouzi IA/CERI, University of Avignon, Agroparc BP 1228, 911 Avignon, France{ralph.elkhoury,

[2] OPHMR: Optimiz Polymorphic Hybrid Multicast Routing Protocol for ad hoc Adel Ben Mnaouer, Member, IEEE, Lei Chen, Student Member, IEEE, Chuan Heng Foh, Member, IEEE, and Juki Wirawan Tantra,

[3] ORRP: Optimal Reactive Routing Protocol for Mobile Ad-Hoc Networks Nabendu Chaki University of Calcutta Calcutta

[4] P.R. China xinming@ustc.edu.cn, {liuqiong,

[5] Interference-Aware QoS OLSR for Mobile Ad-hoc Network Routing Dang-Quan Nguyen, Pascale Minet INRIA, Rocquencourt 78153 Le Chesnay Cedex, France dang-quan.nguyen@inria.fr, pascale.minet@inria.fr

IOP Conf. Series: Materials Science and Engineering **51** (2013) 012026 doi:10.1088/1757-899X/51/1/012026

[6] Interference-Aware QoS OLSRfor Mobile Ad-hoc Network Routing Dang-Quan Nguyen, Pascale Minet INRIA, Rocquencourt 78153 Le Chesnay Cedex, France dang-quan.nguyen@inria.fr, pascale.minet@inria.fr

[7] IdSRF: ID-based Secure Routing Framework for Wireless Ad-Hoc Networks Wei Ren1, Yoohwan Kim1, Ju-Yeon Jo2 1 School of Computer Science, University of Nevada, Las Vegas, NV 89154 {renw, yoohwan, jo, meiyang,

[8] The COMMIT Protocol for Truthful and Cost-Efficient Routing in Ad Hoc Networks with Selfish Nodes, IEEE TRANSACTIONS ON MOBILE COMPUTING, VOL. 7, NO. 1, JANUARY 2008 Stephan Eidenbenz, Giovanni Resta, and Paolo Santi

Comparative study of SVPWM (space vector pulse width modulation) & SPWM (sinusoidal pulse width modulation) based three phase voltage source inverters for variable speed drive

Waheed Ahmed[1,2] and Syed M Usman Ali[1]
[1] Department of Electronic Engineering, NED University of Engineering & Technology, University Road, Karachi -75270, Pakistan.
[2] General Tyre & Rubber Company of Pakistan Limited, H23/2, Landhi Industrial Trading State, Landhi, Karachi, Pakistan.

Engineering_waheed@hotmail.com

Abstract. We have performed comparative studies of Space Vector Pulse Width Modulation (SVPWM) and Sinusoidal Pulse Width Modulation (SPWM) techniques utilizing MATLAB tools. During these investigations, we carried out intensive simulations, comprehensively analyzed the obtained results and compared the harmonic density, power factor (PF), & switching losses of SVPWM and SPWM. It has been observed during investigations that if the switching frequency is high then losses due to harmonics are negligible, thus based on obtained results we suggested that the SVPWM technique is a more reliable solution. Because SVPWM utilizes DC bus voltage more efficiently, generates less Total Harmonic Distortion (THD) and has higher output quality it provides flexible control of output voltage and output frequency for Variable Speed Drive (VSD).

Index Terms: Space Vector Pulse Width Modulation (SVPWM), Sinusoidal Pulse Width Modulation (SPWM), MATLAB, Total Harmonic Distortion (THD), Power Factor (PF), and Variable Speed Drive (VSD) Voltage Source Inverter (VSI).

1. Introduction

Indeed, with the advent of advancement in manufacturing & designing, AC drives super-ceded the DC drives due to their higher performance and better efficiency. The most important advantages of AC drives over DC drives are faster predictable dynamic response, constant and better Power Factor (PF) and 100% continuous torque at zero speed (Applicable to flux vector control technology). AC motors have low cost, less complicated, rugged and easy to maintain or replace, therefore this is also a major cause of shifting of technology from DC drives to AC drives.

For optimum performance, we select the best possible switching and control technique for AC drives. The control techniques used in Variable Speed Drives (VSD) are scalar control (Volts per Hertz Control), vector control (Direct torque control and Field-oriented control) where as the switching techniques used in Variable Speed Drive (VSD) are six step switching technique, selected-harmonic-elimination pulse width modulation (SHPWM), delta pulse width modulation (DPWM), delta-sigma pulse width modulation (DSPWM), minimum-ripple current pulse width modulation and sinusoidal pulse width modulation (SPWM) etc.

The Sinusoidal Pulse Width Modulation and Space Vector Pulse Width Modulation are most widely used modulation techniques due to their greater advantages. The SVPWM technique gives higher level of fundamental voltage as compared to SPWM [1].The comparison of these two modulation gives the results that SVPWM is the best and most reliable modulation because it enables efficient use of DC voltages and smartly works with vector control thus, gives less Total Harmonic Distortion (THD), better PF, and less switching losses at high frequencies.

2. Voltage source inverters

Voltage Source Inverters (VSI) have proven to be more efficient, cost effective, less space, faster dynamic response for rapid changes in speed or torque and be capable of running the motor without de-rating.

A typical three phase VSI consist of a diode rectifier (SCR bridge rectifiers in case of variable DC output voltages) which converts AC line to DC, a parallel capacitor DC Link which stores the energy for the system and regulates the DC bus voltages, an inverter is consist of insulated gate bipolar transistors (IGBTs) which provides variable frequency output depending upon the applied reference voltage and switching technique. A typical VSI is shown in figure 1.

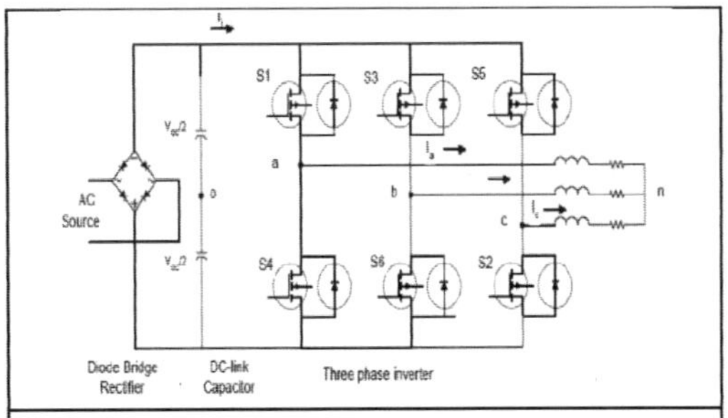

Figure 1. Three phase voltage source inverter

3. Simulink model for SPWM and SVPWM modulation technique

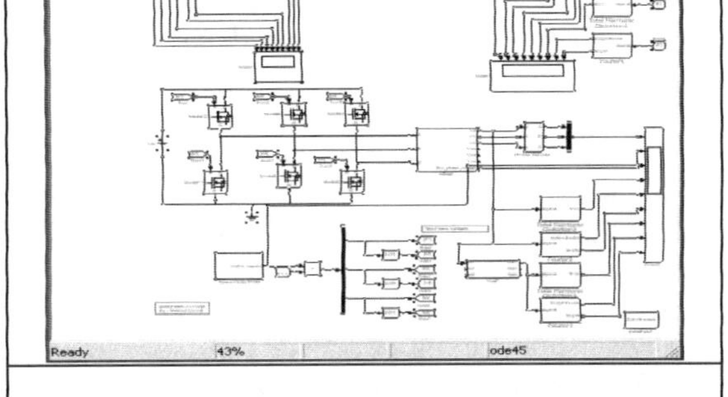

Figure 2. Simulink model for SPWM and SVPWM

4. Sinusoidal pulse width modulation

For the generation of pure sinusoidal signal, SPWM is the most popular technique. In SPWM a digital waveform is generated and the duty cycle is modulated such that the average voltage of the waveform is corresponds to a pure sine wave. SPWM moves the voltage harmonic components to the higher frequencies. The SPWM technique treats each modulating voltage as a separate signal and compared to the common carrier triangular waveform.

The working principle of SPWM includes the following points:

- The frequency of triangular wave is the frequency of PWM.
- Frequency of control voltage controls the fundamental frequency.
- The peak value of control voltage controls the amplitude.
- Modulation Index is defined by:

$$m = \frac{Peak\ of\ V_{ao1}}{V_{d/2}} = \frac{V_{control}}{V_{triangle}}$$

  Where V_{ao1} = Fundamental component of V_{ao}.
- For Phase A:
  If $V_{control} > V_{triangle}$ then $V_{ao} = V_{d/2}$.
  If $V_{control} < V_{triangle}$ then $V_{ao} = -V_{d/2}$.
- For Phase B:
  If $V_{control} > V_{triangle}$ then $V_{bo} = V_{d/2}$.
  If $V_{control} < V_{triangle}$ then $V_{bo} = -V_{d/2}$.
- For Phase C:
  If $V_{control} > V_{triangle}$ then $V_{co} = V_{d/2}$.
  If $V_{control} < V_{triangle}$ then $V_{co} = -V_{d/2}$.

5. Simulation of three phase SPWM based inverter

We analyzed the output line current I_c in MATLAB Simulink and found following readings for THD & Power Factor in table 1:

Table 1. Modulation index and corresponding total harmonic distortion with overall power factor.

Modulation index	THD for current (%)	Overall PF
0.5	137.64	
0.7	60.25	0.829
0.9	5.91	

The ration of the real power to the apparent power is defined by PF. The lesser the phase shift between the fundamental voltage and current, the greater the PF. It should be noted that PF is not directly related to the THD because in THD calculation, phase shift between current and voltage inputs are not factored. However (PF / $Cos\Phi_1$) term can be related to the THD which shows if THD increased then the factor (PF / $Cos\Phi_1$) is decreased. Therefore the readings of power factor in table 1 are measured through scope in MATLAB Simulink instead of calculation by THD and PF formulas. THD is actually the RMS distortion of all AC components excluding fundamental component, THD is expressed as a percentage of the RMS value of the fundamental component. It means THD shows the amount of deviation of the output signal from fundamental signal. Formulas of THD and PF are given in equation (1) and equation (2) respectively.

$$THD = (A_{rms}\text{-}A_1)^{1/2} / A_1 \quad (1)$$

Where A_1 = RMS Value of Fundamental signal.
A_{rms} = Total RMS Value of the complete signal.

$$PF = \mathrm{Cos}\Phi_1 / (1+THD^2)^{1/2} \quad (2)$$

Where Φ_1 = Phase Shift between Fundamental Voltage and Current.

For the simulation, we selected the following parameters parameters to observe the comparison between SPWM and SVPWM techniques:
V_{dc} = 600V
Carrier Signal Frequency = 3K Hz
Fundamental Frequency = 50 Hz
Modulation Index = 0.1 to 1

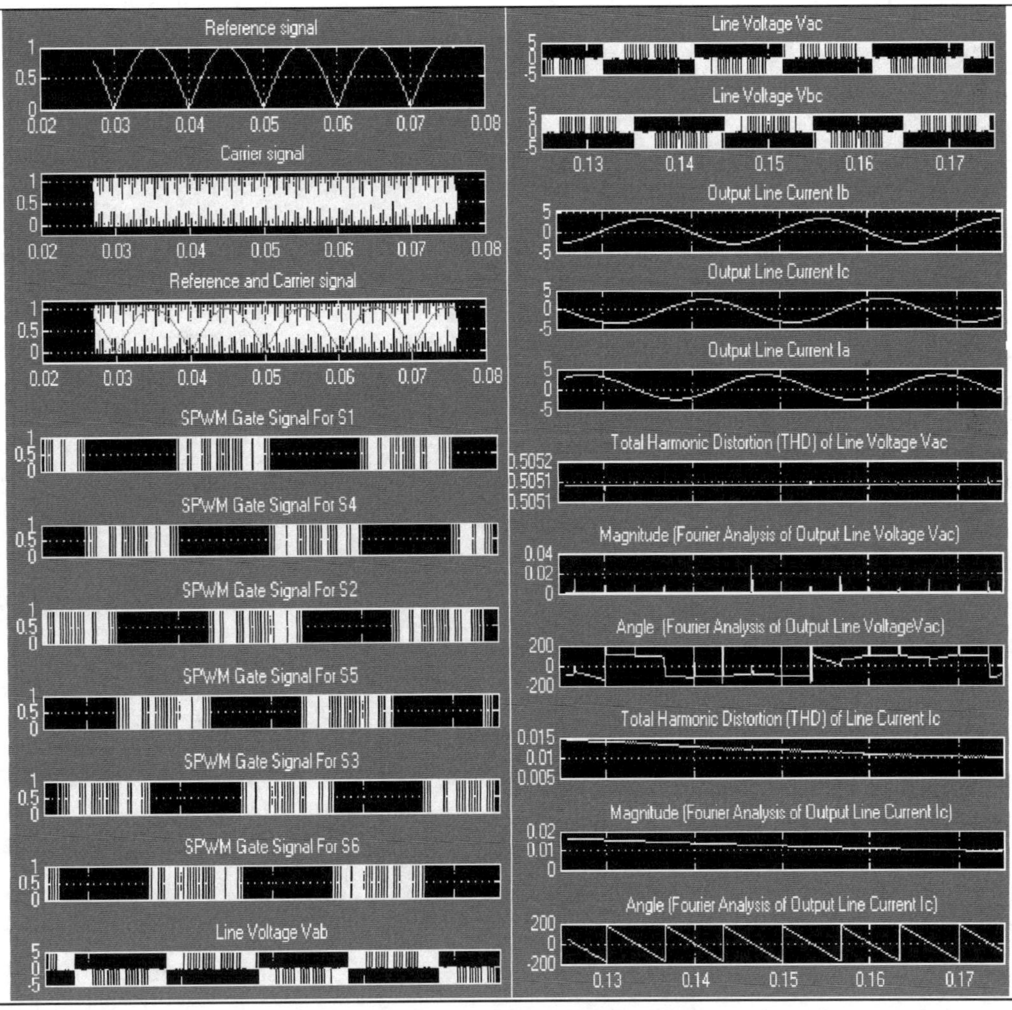

Figure 3. Simulation results of the SPWM

Figure 3 shows the graphical analysis of the reference signal, carrier signal, gate signals for switch S_1 to S_6, output line voltages V_{ab}, V_{ac}, V_{bc}, output line currents I_a, I_c, and I_b respectively. THD of output line voltage V_{ac}, magnitude (Fourier analysis of output line voltage V_{ac}), angle (Fourier analysis of output line voltage V_{ac}), THD of output line current I_c, magnitude (Fourier analysis of output line current I_c), and angle (Fourier analysis of output line current I_c) simultaneously.

Graphical representation of modulation index and THD in SPWM is shown in figure 4 which represents that THD decreases as modulation index increases.

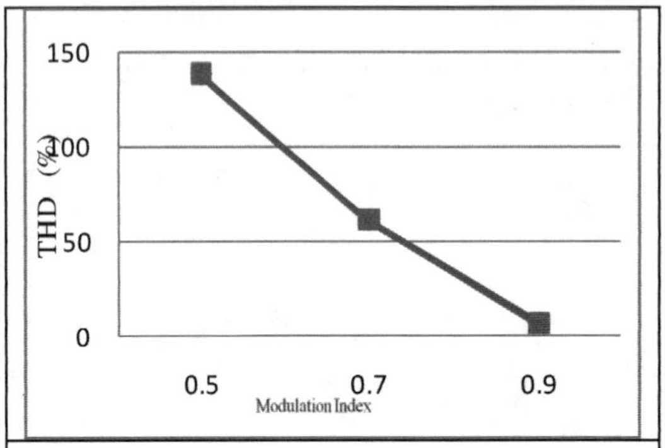

Figure 4. Modulation index vs THD in SPWM

6. Space vector pulse width modulation

SVPWM is the best computational PWM technique for a three phase voltage source inverter because of it provides less THD & better PF. SVPWM works on the principle that when upper transistor is switched ON; corresponding lower transistor is switched OFF. The ON and OFF state of the upper switches (S_1, S_3, S_5) evaluates the output voltages. Switching states and corresponding phase and line voltages are shown in table 2.

Table 2. Switching states and corresponding phase and line voltages.

Switching States			Phase Voltages			Line Voltages		
a	**b**	**c**	V_a	V_b	V_c	V_{ab}	V_{bc}	V_{ca}
0	0	0	0	0	0	0	0	0
1	0	0	2/3	-1/3	-1/3	1	0	-1
1	1	0	1/3	1/3	-2/3	0	1	-1
0	1	0	-1/3	2/3	-1/3	-1	1	0
0	1	1	-2/3	1/3	1/3	-1	0	1
0	0	1	-1/3	-1/3	2/3	0	-1	1
1	0	1	1/3	-2/3	1/3	1	-1	0
1	1	1	0	0	0	0	0	0

For the sector determination, we developed a flow chart showing the different conditions for different sectors. Six sectors and their conditions are shown in figure 5. SVPWM provides approximately 10-20% increment in maximum voltage. The SVPWM can be implemented with the following steps [3] [4]:

- Determination of V_d, V_q, V_{ref}, and angle (α).
- Time duration T_1, T_2, T_0 Determination.
- Determination of the switching time of each transistor (S_1 to S_6).

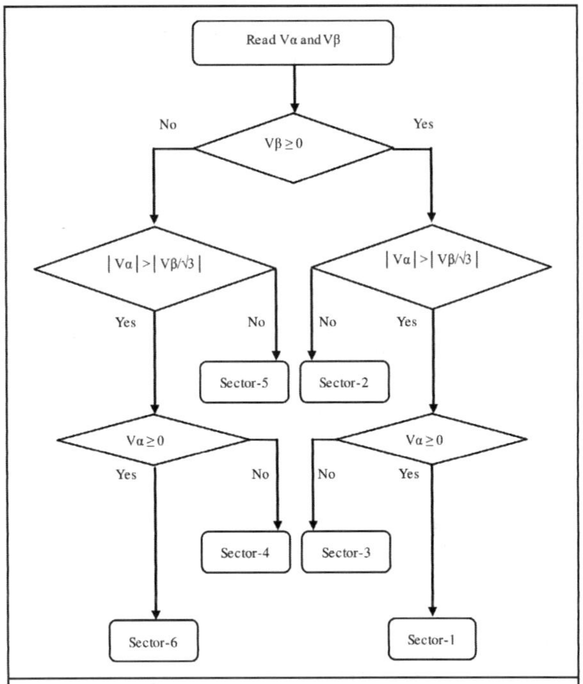

Figure 5. Flow Chart of Sector Determination

7. Simulation of three phase SVPWM based inverter

In SVPWM the overall power factor & THD has improved as compared to SPWM. Analysis of the output line current I_c in MATLAB Simulink for THD & Power Factor is shown in table 3. It should be noted that power factor reading is overall power factor of the system analyzed by power factor analyzer.

Table 3. Modulation index and corresponding total harmonic distortion with overall power factor.

Modulation index	THD for current (%)	Overall PF
0.5	129.62	
0.7	40.71	0.871
0.9	3.42	

In the simulation of SVPWM, we selected the following parameters to observe the comparison between SPWM and SVPWM techniques:
V_{dc} = 600V
Carrier Signal Frequency = 3K Hz
Fundamental Frequency = 50 Hz
Modulation Index = 0.1 to 1

Figure 6 is showing the graphical results of three phase output line voltages V_{ab}, V_{bc}, V_{ac}, transformed two phase voltages, control signals of SVPWM, THD of output line voltage V_{ac}, magnitude (Fourier analysis of output line voltage V_{ac}), angle (Fourier analysis of output line voltage V_{ac}), THD of output line current I_c, magnitude (Fourier analysis of output line current I_c), and angle (Fourier analysis of output line current I_c) simultaneously. SVPWM has good drive response if we have variable speed reference or load torque [2]. Graphical representation of modulation index & THD in SVPWM are also shown in figure 6.

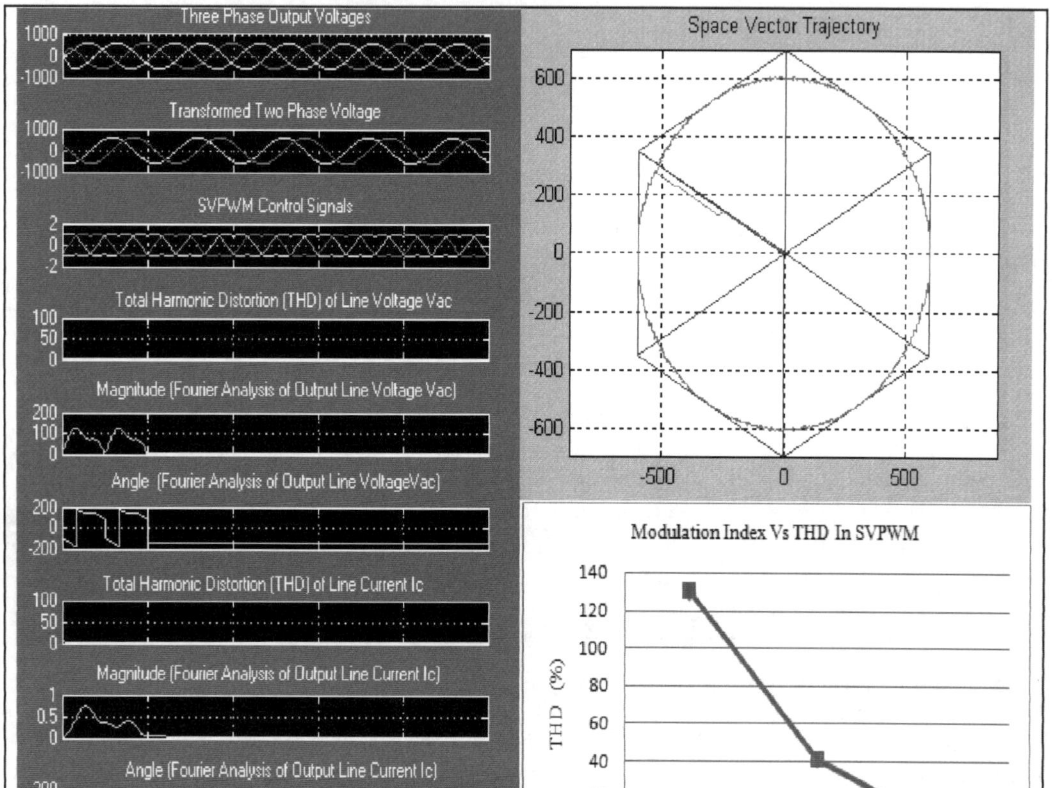

Figure 6. Simulation results of the SVPWM.

8. Switching losses in SPWM and SVPWM

For low power applications, switching losses are acceptable for specific range but for high power applications, switching losses become more significant. Because of switching losses, high frequencies (greater than 20 kHz) are less efficient than lower frequencies (as low as 100 Hz) due to efficiency of system reduces as switching losses increased, since for reducing filtering requirements we have to increase switching frequency which results in greater switching losses. Although switching losses can be reduced by modifying carrier signal in SPWM or using zero switching technique or shifting to multilevel inverters [5] but on the other hand it results in greater harmonic distortion or poor power factor.

SVPMW has greater flexibility to reduce switching losses. In SVPWM reduced switching losses are because of the changing of any one switching state which results in one single phase voltage change every time. If system needs further reduction in switching losses than another technique could be used for switching loss reduction based on stopping the control pulses of SVPWM for some duration and this duration depends upon angle of the load power factor. For different modulation indexes, extra switching can be eliminated in SVPWM.

9. Conclusion

In this paper, SPWM and SVPWM techniques have been investigated and compared. During the investigations, we realized harmonic density, power factor & switching losses in both techniques. For this purpose we also performed extensive simulations of these techniques using MATLAB tools. It has been observed that SVPWM has showed superior performances due to less THD, greater PF and less switching losses because SVPWM utilizes advance computational switching technique to reduce THD. It also reduces switching losses because of the changing of any one switching state which results in one single phase voltage change every time. Furthermore, at high switching frequencies SVPWM gives better results as compared to SPWM. Thus, based on all obtained results, we concluded that SVPWM technique provides greater overall performance and efficiency as compared to SPWM technique.

References

[1] M.SajediHir, Y. Hoseynpoor, P. MosadeghArdabili, T. PirzadehAshraf 2011Modeling and Simulation of Induction Motor Drive with Space Vector Control *Australian Journal of Basic and Applied Sciences* **5(9): 2210-2216** 1

[2] Preeti Soni and Kavita Burse 2012 Analysis of Voltage Source Inverters using Space Vector PWM for Induction Motor Drive *IOSR Journal of Electrical and Electronics Engineering* **ISSN 2278-1676 Vol 2** 19

[3] Ashish Gupta and Sanjiv Kumar 2012 Analysis of Three Phase Space Vector PWM Voltage Source Inverter for ASD's *International Journal of Emerging Technology and Advanced Engineering* **ISSN 2250-2459 Vol 2** 1

[4] K Vinoth Kumar, Prawin Angel Michael, Joseph P John and Dr S Suresh Kumar 2010 Simulation and Comparision of SPWM and SVPWM Control for Three Phase Inverter *ARPN Journal of Engineering and Applied Sciences* **ISSN 1819-6608 Vol 5**

[5] Nguyen Van Nho A Graphical Approach to Switching Losses and Harmonics Distortion For Carrier SVPWM Methods In Multilevel Inverters (HCMUT Vietnam: Dept. of Electrical Engineering, Hochiminh City University of Technology).

Performance Analysis of Blind Beamforming Algorithms in Adaptive Antenna Array in Rayleigh Fading Channel Model

M Yasin[1], Pervez Akhtar[1], Amir Hassan Pathan[2]
[1]National University of Sciences and Technology, Islamabad, Pakistan
[2]Information Systems & Technology Department, State Bank of Pakistan
E-mail: myasin@pnec.edu.pk,

Abstract. In this paper, we analyze the performance of adaptive blind algorithms -- i.e. Kaiser Constant Modulus Algorithm (KCMA), Hamming CMA (HAMCMA) -- with CMA in a wireless cellular communication system using digital modulation technique. These blind algorithms are used in digital signal processor of adaptive antenna to make it smart and change weights of the antenna array system dynamically. The simulation results revealed that KCMA and HAMCMA provide minimum mean square error (MSE) with 1.247 dB and 1.077 dB antenna gain enhancement, 75% reduction in bit error rate (BER) respectively over that of CMA. Therefore, KCMA and HAMCMA algorithms give a cost effective solution for a communication system.

1. Introduction

In any communication system, the aim is to reduce interference level to get better quality of service [1]. This can be possible with digital modulation techniques [2] to make the system more spectrally efficient using beamforming concept [3]. Beamforming is either switch or adaptive. Adaptive one is implemented on signal processor with antenna array to make the antenna smart [4]. The windowing techniques have been well addressed in array signal processing to reduce the sidelobes in [5]. In [6] three different versions of the CMA are introduced, each one motivated by a different signal processing problem. CMA originally employs an FIR filter which accepts complex input data having complex coefficients [7]. Paper [8] addresses CMA algorithms, their analytical work regarding rate of convergence and behaviour. It also scrutinizes the "capture" problem identified in [7]. In [9], a CM Array and its extension are presented. An analytic method for solving the CM problem is described in [10]. In this article, we analyzed the performance of KCMA and HAMCMA with CMA in a communication system which provides optimum results for adaptive antenna array system. This research is the extension of our previous work reported in [11]. This paper investigates beamforming with windowed CMA (i.e. KCMA and HAMCMA) in a cellular communication system and compared with CMA in terms of directive gain, MSE reduction, BER analysis and convergence.

The next section describes the system model. Section III displays a mathematical model for blinds algorithms. Section IV presents simulation results. Discussion and comments is presented in section V. Finally, section VI concludes the paper.

2. System Model

Consider a wireless cellular communication model as shown in Fig 1. A random digital signal is modulated using 16 point QAM technique. When this modulated signal is passed through Rayleigh fading channel, it becomes faded. As a result, echoes of the modulated signal is produced which arrive at the array of antenna system. This multipath fading phenomena cause fluctuations in the received signal's amplitude, phase, and angle of arrival. Then received signal is processed through these blind beamforming algorithms one by one. The received signals are filtered by weights of antenna elements. The outputs of each antenna elements are coupled to the summer and generate the array output. The weights of antenna arrays determine the magnitude and phase of signals which helps in beam formation. As a result optimum beam towards desired direction is achieved and minimum energy is transmitted towards interferers, thus creating null towards them. The output of beamformer is demodulated and computes the BER. The output of demodulator has noise parameters and is passed through an affine projection (AP) adaptive filter that extracts desired signal.

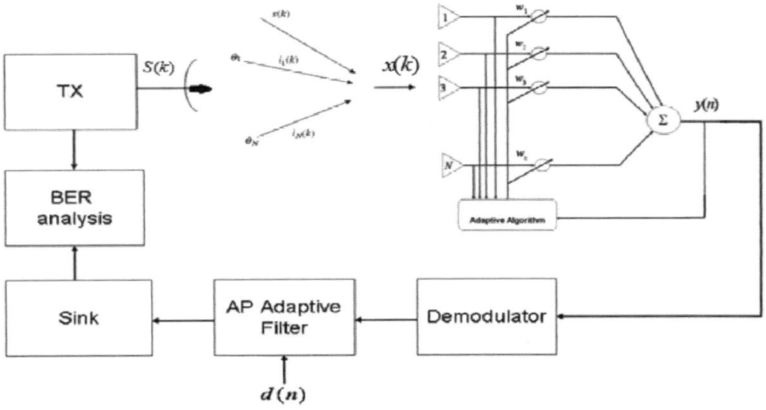

Fig 1. Implementation of blind adaptive algorithm in Rayleigh Fading Channel Model

3. Mathematical Model

3.1. Kaiser Constant Modulus Algorithm

The KCMA algorithm is the unification of CMA and Kaiser Window. Now consider a linear beamformer employing window technique method as shown in Fig. 1. Then its output will be

$$y_k = \mathbf{X}_k (\hat{\mathbf{W}}_k)^{\mathrm{T}} \tag{1}$$

where $\hat{\mathbf{W}}_k = kaiser(N,\beta)\mathbf{W}_k$ is the initial estimate weight vector using Kaiser window. β is the Kaiser window parameter that affects the sidelobes attenuation and N represents the number of elements in an array. The signal array vector having different amplitudes and phases is written by

$$\mathbf{X}_k = [x_1, x_2, \ldots\ldots, x_N]^T \tag{2}$$

The error signal for blind algorithm is given by

$$e_k = y_k - \frac{y_k}{|y_k|} \tag{3}$$

Putting value of y_k in (3) and differentiate w.r.t. weight $\mathbf{W}$, then we get

$$\frac{\partial e_k}{\partial \mathbf{W}} = [\mathbf{X}_k kaiser(N,\beta)] = \mathbf{X}_k kaiser(N,\beta) \tag{4}$$

Putting this value in the gradient estimate of the form giving by

$$\hat{\nabla}_k = 2e_k \begin{bmatrix} \frac{\partial e_k}{\partial \mathbf{W}_0} \\ . \\ . \\ \frac{\partial e_k}{\partial \mathbf{W}_L} \end{bmatrix} = 2e_k (\mathbf{X}_k kaiser(N,\beta)) \tag{5}$$

Mathematically, the gradient decent rule [12 chap 2 (2.35) and 4 (4.36)] is given by

$$\mathbf{W}_{k+1} = \mathbf{W}_k - \mu \hat{\nabla}_k \tag{6}$$

Putting value of gradient estimate (5) into (6), we get

$$\mathbf{W}_{k+1} = \mathbf{W}_k - 2\mu e_k kaiser(N,\beta)\mathbf{X}_k \tag{7}$$

This is the required weight vector for KCMA using Kaiser Window, where μ is the step size.

3.2. *Hamming Constant Modulus Algorithm*

This proposed algorithm [11] is designed using Hamming window technique. The weight vector for HAMCMA can be derived on the same pattern as KCMA and is given by

$$\mathbf{W}_{k+1} = \mathbf{W}_k - 2\mu e_k hamming(N)\mathbf{X}_k \tag{8}$$

The array output signal obtained with the sample weights is given by

$$y_k = \mathbf{X}_k (\hat{\mathbf{W}}_k)^{\mathbf{T}} \tag{9}$$

where $\hat{\mathbf{W}}_k = hamming(N)\mathbf{W}_k$ is the initial estimate weight vector using Hamming window function and $\mathbf{X}_k$ is the input samples. The weight matrix update (8) approaches its true value when the number of samples grows.

3.3. *Constant Modulus Algorithm*

CMA is a blind algorithm based on the idea to maintain gain on the signal while minimizing the total output energy. The weight update equation of CMA algorithm [4] is given by

$$\mathbf{W}_{k+1} = \mathbf{W}_k - 2\mu e_k \mathbf{X}_k \tag{10}$$

where e_k is the cost function, using (3) and μ is the step size. The array output is given by

$$y_k = \mathbf{X}_k(\mathbf{W}_k)^{\mathrm{T}} \tag{11}$$

3.4. Affine Projection Adaptive Filter

The Affine Projection filter is an adaptive filter that has fast convergence property. It is employed to extract a desired signal from a noise-corrupted signal by filtering out the noise. The adaptive filter produces an output which is a true copy of input desired signal.

4. Simulation Results

The QAM signal (16 numbers of bits per symbol) is applied for simulation purpose, to illustrate the effect of various parameters of digital communication system as thrash out in incoming paragraphs:

4.1. Role of Smart Adaptive Antenna in Communication System under Study

Uniform linear array is taken for adaptive antenna array system with N = 10 and the distance between two elements is supposed to be as $\lambda/2$. 100 samples are taken for simulation purpose. The angle of arrival (AOA) for desired user is 0 degree. The array factor plot is shown in Fig. 2 for KCMA, CMA and HAMCMA that shows one desired signal arriving from the angle 0 degree and two interferer arriving from angles other than desired angle. The ratio of main lobe to the first side lobe is 4.48 dB, 4.46 dB and 3.63 dB for CMA, KCMA and HAMCMA respectively. It is proved that KCMA is giving 1.247 dB gain improvements than that of CMA and at the same time HAMCMA provides 1.077 dB improvements as compared to CMA. However, the null depth performances of all these algorithms under study are same and cancel the interference by placing a null at the angle of interferer.

4.2. Mean Square Error Performance

The minimum MSE describes the performance of the given system. The minimum MSE for three blind beamforming algorithms are shown in Fig. 3 which indicates that KCMA and HAMCMA have minimum MSE (almost zero) at 5th and at 7th iterations as compared to CMA. It is worth mentioning that minimum MSE for these three blind beamforming algorithms changes and effected by number of iterations. Therefore, performance of KCMA and HAMCMA is better over that of CMA.

4.3. Bit Error Rate Performance

In Fig 4, the theoretical BER curve is taken as a reference value for comparison. The plots as shown in Fig 4 describe that the error rates obtained are smaller as compared to theoretical BER; therefore, the BER performance is greatly improved. The computed BER values for HAMCMA, KCMA and CMA algorithms at SNR=9 are 0.001, 0.001 and 0.004 respectively. The BER values for HAMCMA and KCMA are 25% (0. 001 is 25% of 0. 004) and 25% (0. 001 is 25% of 0. 004) as compared to CMA. Then the BER is reduced by 75% by both HAMCMA and by KCMA. Table compares various results obtained from Fig. 2, 3 and 4 for blind beamforming algorithms under study.

IOP Conf. Series: Materials Science and Engineering **51** (2013) 012028 doi:10.1088/1757-899X/51/1/012028

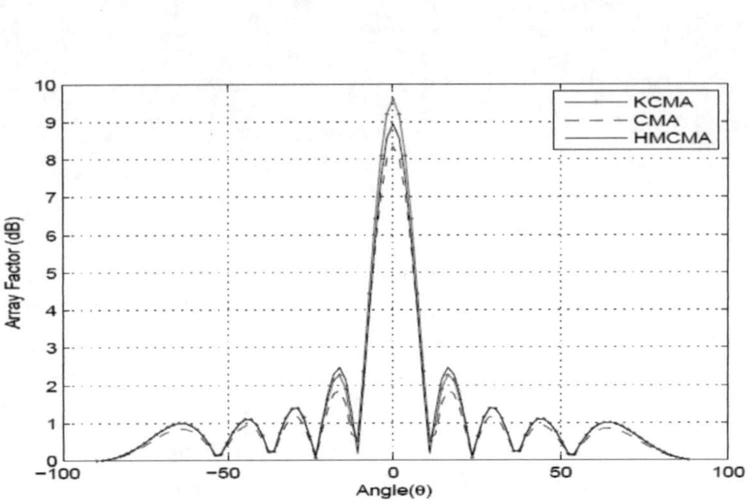

Fig 2. Array factor versus AOA for **desired user**

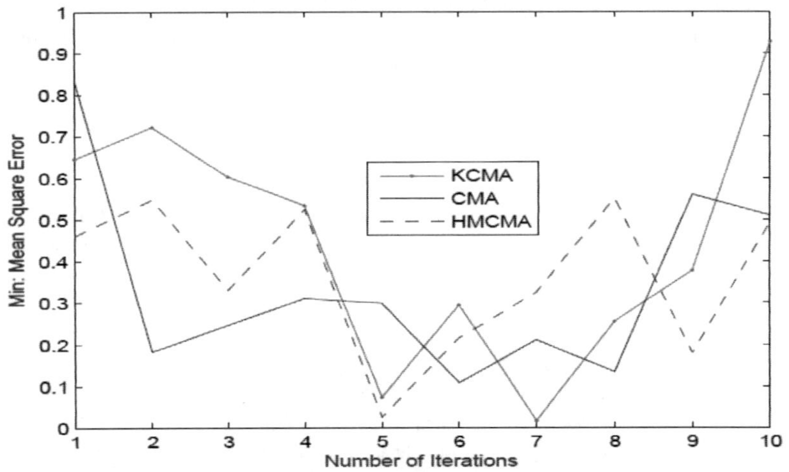

Fig 3. Mean Square Error performance

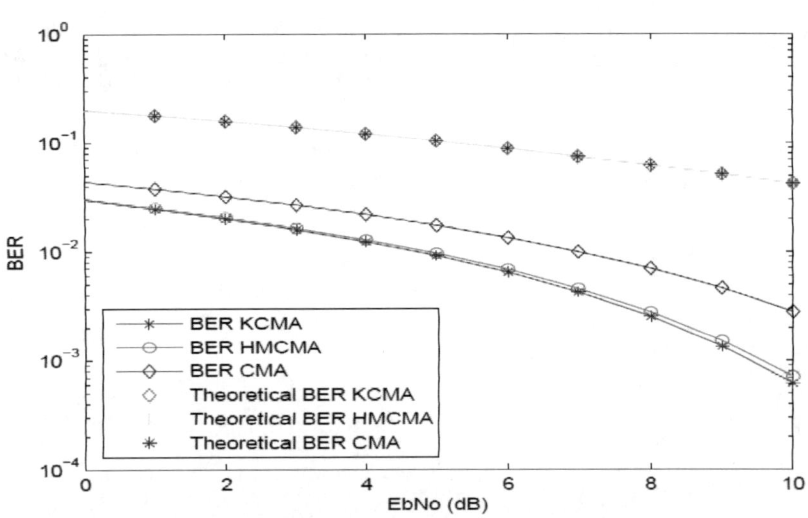

Fig 4. Comparison between Calculated and Theoretical BER

5. Discussion and Comments

The findings of simulation are:

A. The directional gain of the KCMA beamforming algorithm is more as compared to HAMCMA and CMA.
B. The proposed algorithms save power because a directional gain of proposed model is optimum.
C. KCMA and HAMCMA converge at 5th and at 7th iterations respectively.
D. KCMA is giving 1.247 dB whereas HAMCMA provides 1.077 dB gain improvement than that of CMA.
E. KCMA and HAMCMA reduce BER by 75% as compared to CMA.
F. The convergence rate of HAMCMA is smaller than that of CMA and KCMA as given in Table.

Table. System Input and Throughput Estimate for Blind Algorithms under study

Input Parameters	Algorithms		
	KCMA	HAMCMA	CMA
No. of Samples	100	100	100
AOA (degree)	0	0	0
Element Spacing (d)	0.5λ	0.5λ	0.5λ
No. of Elements	10	10	10
Step Size μ	0.05	0.05	0.05
Output Parameters	System Throughput Estimate		
Beam width (degree)	20	20	20
Array Gain (dB)	9.589	8.99	8.342
Sidelobe Level (dB)	2.15	2.47	1.86
BER at 9 dB SNR	0.001	0.001	0.004
Convergence rate (Sec)	1.0299	0.0926	0.1211

6. Conclusion

With respect to CMA, KCMA and HAMCMA provide optimum performance. Therefore, these algorithms can be implemented at base station of mobile communication systems as one of the option.

References

[1] C. Alor and D. H. Covarrubias, (2003). Optimization on RF Devices Design Applied in Wireless Communications. *Journal of Applied Research and Technology*, Vol. **1**, No. 1, 27-39.

[2] Shuang Tan, Sheng Chen & Lajos Hanzo, (2008). Iterative Multiuser Minimum Symbol Error Rate Beamforming Aided QAM Receiver. *IEEE Signal Processing Letters*, Vol. **15**, 301-304.

[3] Daniel J. Ryan, I. Vaughan L. Clarkson, Iain B. Collings et al. (2009). 'AM and PSK Codebooks for Limited Feedback MIMO Beamforming. *IEEE Transactions on Communications*, Vol. **57**, 1184-1196.

[4] Lal. C. Godara, (1997). Applications of Antenna Arrays to Mobile Communications, Part II; Beam-Forming and Directional of Arrival Considerations. *Proceeding of the IEEE*, Vol. **85**, No. 8, 1195-1245.

[5] Harry L. Van Trees., (2002). Detection, Estimation, and Modulation Theory, Part IV, Optimum Array Processing. Wiley.

[6] J.R. Treichler and M.G. Larimore, (1985). New processing techniques based on constant modulus adaptive algorithm. *IEEE Trans. Acoust., Speech, Signal Processing*, Vol. **33**, 420–431.

[7] J. R. Treichler and B. G. Agee, (1983). A new approach to multipath correction of constant modulus signals. *ZEEE Trans. Acoust., Speech, Signal Processing,* Vol. ASSP-**31**, 349-472.

[8] J. R. Treichler and M. G. Larimore, (1985). The tone capture properties of cma-based interference suppressors. *IEEE Trans. on Acoust., Speech and Signal Processing*, ASSP-**33(4)**, 946-958.

[9] R. Gooch and J. Lundell, (1986). The CM array: An adaptive beamformer for constant modulus signals. *in Proc. IEEE ICASSP*, (Tokyo), 2523–2526.

[10] Alle-Jan van der Veen and Arogyaswami Paulraj, (1996). An Analytical Constant Modulus Algorithm. *IEEE Transactions on Signal Processing*, Vol. **44**, No. 5, 1-19.

[11] M Yasin, Pervez Akhtar, S.H Zaheer Naqvi, (2011). Design and Performance Analysis of Blind Algorithms for Smart Antenna System Using Window techniques. *Middle East Journal of Scientific Research* **7** (**4**), 458-466.

[12] B. Widrow and S.D. Steams, (1985). Adaptive Signal Processing. Prentice-Hall Inc., New Jersey.

Cloud vector mapping using MODIS 09 Climate Modeling Grid (CMG) for the year 2010 and 2011

Asjad Asif Jah[1], Yousaf Bin Farrukh[2] and Rao Muhammad Saeed Ali[3]
[1-3] Institute of Geographical Information Systems, (SCEE),
National University of Sciences and Technology, Islamabad, Pakistan

E-mail: asjadjah@gmail.com

Abstract. An alternate use for MODIS images was sought by mapping cloud movement directions and dissipation time during the 2010 and 2011 floods. MODIS Level-02 daily CMG (Climate Modelling Grid) land-cover images were downloaded and subsequently rectified and clipped to the study area. These images were then put together to observe the direction of cloud movement and vectorize the observed paths. Initial findings suggest that usually cloud does not have a prolonged coverage period over the northern humid region of the country and dissipates within less than 24-hours. Additionally, this led to the development of a robust methodology for cloud motion analysis using FOSS and market leading GIS utilities.

Keywords: Satellite Remote Sensing, Cloud mapping, Climate Modelling Grid, Pakistan, Floods

1. Introduction

Pakistan is situated in a region with a sub-season in summer called the monsoon. The monsoon of 2010 and 2011 were special due to the fact that they caused widespread devastation across wide swaths of land in the country. In the floods of 2010 more than 20% of the territory was inundated alone [1] and similarly the 2011 floods caused havoc in Southeastern Sindh province.

During the monsoon overcast conditions prevails in various parts of the country. An analysis of these conditions and their correlation with atmospheric pollutions, the urban heat island phenomena and rainfall distribution is expected to deliver interesting scientific knowledge. In this regard a mapping exercise using the MODIS, MOD09 CMG product (see figure 1) was sought using Open Source GDAL (Geospatial Data Abstraction Library) and ArcGIS. A series of images were downloaded for the monsoon season in 2010 and 2011 from the NASA LAADSWEB data portal. These images were then clipped and converted to RGB using band 1, 4 and 3 and subsequently used for mapping cloud movement and dissipation durations.

Figure 1: MODIS Climate Modeling Grid RGB (1,4,3) [2]

2. Data and methods

A selected set of MOD09 CMG (Level-2 climate modelling grid) product tiles were downloaded from the NASA LAADS web data portal for the months of July and Aug for the years 2010 and 2011. These images were stored on a dedicated cluster meant for MODIS low to high resolution imagery processing due to the immense data size.

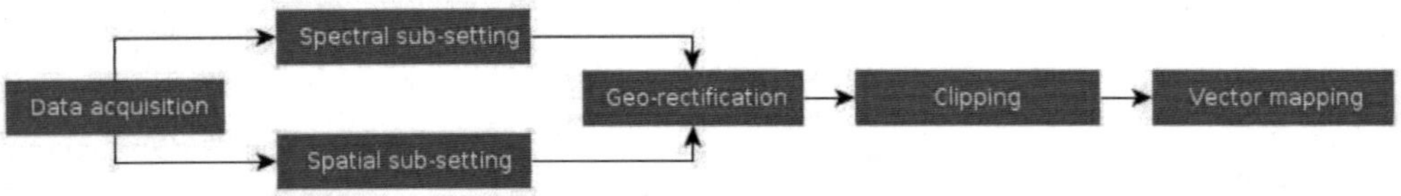

Figure 2 : Flow diagram of the steps that were involved in this analysis.

```
# Band Composite On All Files in a Folder

import arcpy
from arcpy import env

arcpy.env.workspace = "G:/files"
arcpy.env.overwriteOutput = True

files = arcpy.ListFiles("*.tif")

# Provide the total number of files in the folder
totalFiles = XX

#Give total bands per scene
totalBands = X

i=0
outFile=1

while i+2 < totalFiles:
       # Change the order of the three files as per RGB
```

```
        print("Band Composite Processing   " + files[i] + " + " + " + files[i+1]
+ " + " + files[i+2])
        #output file will be numbered in acsending order, you can also
prefix an identifier in the output name
        arcpy.CompositeBands_management(files[i]+ ";" +files[i+1]+ ";" +
files[i+2],"G:/out/" +str(outFile)+".tif")
        i=i+totalBands
        outFile = outFile+1
```

Script 1: Python script used in ArcGIS for band composite generation

The MODIS CMG product tiles bear a coarse resolution of 0.05° and cover the entire globe on a regular basis. For this exercise the choice was that of the former due to limited processing capacity and significantly high data size of ~130 GB. The MODIS (MOD09 CMG) data is distributed in HDF-4 format and hence contains a set of 7 surface reflectance sub-dataset bands of which band 1, 4 and 3 are suitable for RGB generation, in respective order [2].

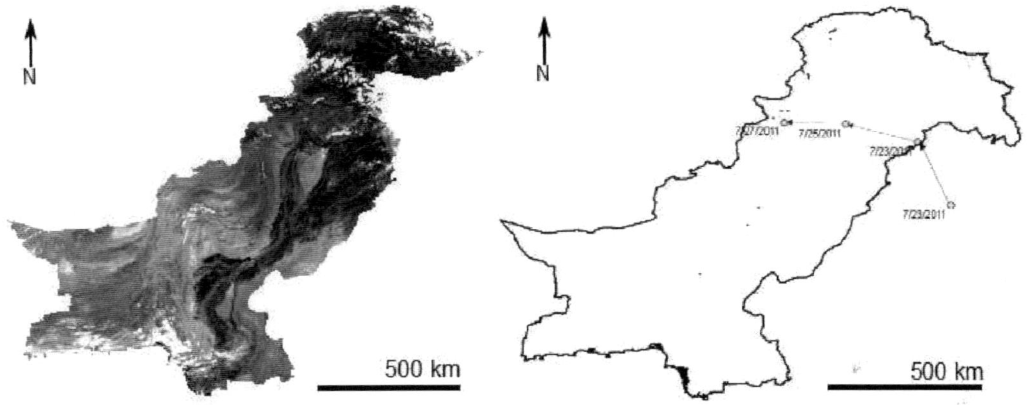

Figure 3: A sample RGB for created using the acquired MOD09 images and a cloud movement map generated from a series of similar MOD09 images.

A custom bash script was written to spectrally and spatially subset these files which were 116 in number and convert these bands to TIFF format. Once this was done these images were imported using another bash script in GRASS-GIS [3] and clipped to the political boundaries of Pakistan. RGBs were developed for our vectorization needs wherever required using a custom python script in ArcGIS (see script 1), and clouds and their direction of advance was mapped accordingly (see figure 2 and figure 3).

3. Results and findings

This cloud mapping exercise was critical to our understanding of the horizontal spread of clouds across Pakistan, their development and dissipation rates and forms over different parts of the country. With such clouds maps we plan to further incorporate precipitation records from the TRMM (Tropical Rainfall Measurement Mission) and develop cross-tabulation maps for cloud types and precipitation variability over the entire territory.

4. Conclusion

The cloud distribution maps helped us observe an extended residence period of two days over the northern parts of the country. This forms a humid belt of high rainfall in the Hindu-Kush, Karakoram and Himalayan piedmont. Moreover, such persistent cloud cover was rare and mostly the cloud would dissipate in less than 24 hours. FOSS-GIS coupled with ArcGIS is a handy utility for mapping decadal time-series variations in cloud cover as this work proves and their shift in any seasonal variations.

IOP Conf. Series: Materials Science and Engineering **51** (2013) 012029 doi:10.1088/1757-899X/51/1/012029

5. References

[1] Robotka, B. Pakistani Floods 2010: An analysis of its impact on the country and its people, 2011.

[2] E.Vermote and J. Ray. MODIS surface reflectance user's guide, 2011.

[3] Neteler M., Bowman M.H., et.al. GRASS GIS: A multi-purpose Open-Source GIS. Environmental Modelling & Software, 2012.

6. Acknowledgements

We would like to thank the NUST IGIS faculty and Mr. Kamran Aslam for their kind support and guidance.

AUTHOR INDEX